[illegible] physic[illegible]
at the U[illegible]nchester as well as researcher on [illegible] the most ambitious experiments on Earth, the ATLAS experiment on the Large Hadron Collider in Switzerland. He is best known to the public as a science broadcaster and presenter of the popular BBC *Wonders* trilogy.

Andrew Cohen is Head of the BBC Science Unit and the Executive Producer of the BBC series *Forces of Nature*. He has been responsible for a wide range of science documentaries including *Horizon*, the *Wonders* trilogy, *Human Universe* and *Stargazing Live*. He lives in London with his wife and three children.

Praise for Professor Brian Cox:

'Engaging, ambitious and creative.'
— *Guardian*

'He bridges the gap between our childish sense of wonder and a rather more professional grasp of the scale of things.'
— *Independent*

'If you didn't utter a wow watching the TV, you will while reading the book.'
— *The Times*

'In this book of the acclaimed BBC2 TV series, Professor Cox shows us the cosmos as we have never seen it before – a place full of the most bizarre and powerful natural phenomena.'
— *Sunday Express*

'Cox's romantic, lyrical approach to astrophysics all adds up to an experience that feels less like homework and more like having a story told to you. A really good story, too.'
— *Guardian*

'Will entertain an[illegible]at would be.'
— *Independ[illegible]*

For my dad, David.
— Brian Cox

For Benjamin, Martha, Theo, Dan, Jake, Lyla, Ellie, Toby, Phoebe, Max, Zak, Josh, Isaac and Tabitha because curious young minds always ask the smartest of questions.
— Andrew Cohen

FORCES OF NATURE

PROFESSOR
BRIAN COX
& ANDREW COHEN

William Collins
An imprint of HarperCollinsPublishers
1 London Bridge Street
London SE1 9GF
WilliamCollinsBooks.com

This William Collins paperback edition published in 2017

22 21 20 19 18 17
10 9 8 7 6 5 4 3 2 1

First published in the United Kingdom by William Collins in 2016

Typeset and designed in Graphik and Caslon by Zoë Bather

By arrangement with the BBC.

A catalogue record for this book is available from the British Library.

ISBN 978-0-00-821003-8

Printed and bound in Great Britain by Clays Ltd, St Ives plc.

SEARCHING FOR THE DEEPEST ANSWERS

TO THE SIMPLEST QUESTIONS

'What beauty. I saw clouds and their light shadows on the distant dear Earth… The water looked like darkish, slightly gleaming spots… When I watched the horizon, I saw the abrupt, contrasting transition from the Earth's light-coloured surface to the absolutely black sky. I enjoyed the rich colour spectrum of the Earth. It is surrounded by a light blue aureole that gradually darkens, becoming turquoise, dark blue, violet, and finally coal black.'
— Yuri Gagarin

Taking a different perspective

This is a book about science. What is science? That's a good question, and there may be as many answers as there are scientists. I would say that science is an attempt to understand the natural world. The explanations we discover can often seem abstract and separate from the familiar, but this is a false impression. Science is about explaining the everyday minutiae of human experience. Why is the sky blue? Why are stars and planets round? Why does the world keep on turning? Why are plants green? These are questions a child might ask, but they are certainly not childish; they generate a chain of answers that ultimately lead to the edge of our understanding.

If you dig deep enough, most questions end with uncertainty. The sky is blue because of the way light interacts with matter, and the way light interacts with matter is determined by symmetries that constrain the laws of Nature. We'll encounter these concepts later in the book. But if one keeps on digging, and asks why those particular symmetries, or why there are laws of Nature at all, then we are into the glorious hazy place in which scientists live and work; the space between the known and the unknown. This is the domain of the research scientist, and it is a place of curiosity and wonder.

Grander questions lurk in the half-light. How did life on Earth begin? Is there life on other worlds? What happened in the first few moments after the Big Bang? These are questions that have a sense of depth and a feeling of complexity and intractability, but the techniques and processes by which we look for answers are no different to those deployed in discovering why the sky is blue. This is an important point. If a question sounds deep, it doesn't mean that the way to answer it is to retire to the wilderness for a year, sit cross-legged and hope for something to occur to you. Rather, the answers

'THE FIRST DAY OR SO WE ALL POINTED TO OUR COUNTRIES. THE THIRD OR FOURTH DAY WE WERE POINTING TO OUR CONTINENTS. BY THE FIFTH DAY WE WERE AWARE OF ONLY ONE EARTH.'

— SULTAN BIN SALMAN BIN ABDULAZIZ AL-SAUD,
SPACE SHUTTLE STS-51-G

'ODDLY ENOUGH
THE OVERRIDING SENSATION
I GOT LOOKING AT THE
EARTH WAS, MY GOD
THAT LITTLE THING IS SO
FRAGILE OUT THERE.'

– MIKE COLLINS, GEMINI 10, APOLLO 11

are often constructed on foundations generated by the systematic and careful exploration of simpler questions. This idea is central to our book. In seeking to understand the everyday world – the colours, structure, behaviour and history of our home – we develop the knowledge and techniques necessary to step beyond the everyday and approach the Universe beyond.

Planet Earth is the easiest place in the Universe to study because we live on it, but it is also confusingly complicated. For one thing, it's the only planet we know of that supports life. It is home to over seven billion humans and at least ten million species of animals and plants. Of its surface area, 29 per cent is land, and humans have divided that 148,326,000 square kilometres into 196 countries, although this number is disputed. Within these boundaries, reflecting the vagaries of 10,000 years of human history, there are over 4000 religions. Some want to increase the number of countries; others want to decrease the number of religions. For such a small world orbiting an ordinary star in such a run-of-the-mill galaxy, it's not very well organised and difficult to understand through the parochial fog. Just over five hundred humans have travelled high enough to see our home from space – a small world against the backdrop of the stars – and when they do, something interesting happens. They see through the fog, and return with a description not of segregation and complexity, but of unity and simplicity.

'When you're finally up at the Moon looking back on Earth, all those differences and nationalistic traits are pretty well going to blend, and you're going to get a concept that maybe this really is one world and why the hell can't we learn to live together like decent people.'
Frank Borman, Gemini 7, Apollo 8

'If somebody had said before the flight, "Are you going to get carried away looking at the Earth from the Moon?" I would have said, "No, no way." But yet when I first looked back at the Earth, standing on the Moon, I cried.'
Alan Shepard, Mercury 3, Apollo 14

The astronauts were not making whimsical comments. These are statements from human beings whose experience has given them a different perspective. The astronauts see simplicity because they have been forced to look at the world in a different way. We are self-evidently one species, inhabiting one planet, and it follows that we have one chance not to mess it all up. We can't all be astronauts, but we can all be scientists, and I think science provides a similar perspective to altitude. It lifts us up, mentally rather than physically, and allows us to survey the landscape below. We look for regularities and, once glimpsed, we try to understand their origin. On his return from space, Scott Carpenter, officer in the United States Navy and Korean War veteran, felt that our highest loyalty should not be to our own country, but to the family of man and the planet at large. Space travel is about a shift in perspective, and so is science. The more we understand about Nature, the more beautiful it appears and the more we understand what a privilege it is to be able to spend our short time exploring it. Be a child. Pay attention to small things. Don't be led by prejudice. Take nobody's word for anything. Observe and think. Ask simple questions. Seek simple answers. That's what we'll do in this book, and hopefully, by the end, you'll agree with Scott Carpenter.

'THIS PLANET IS NOT TERRA FIRMA. IT IS A DELICATE FLOWER AND IT MUST BE CARED FOR. IT'S LONELY. IT'S SMALL. IT'S ISOLATED, AND THERE IS NO RESUPPLY. AND WE ARE MISTREATING IT. CLEARLY, THE HIGHEST LOYALTY WE SHOULD HAVE IS NOT TO OUR OWN COUNTRY OR OUR OWN RELIGION OR OUR HOME TOWN OR EVEN TO OURSELVES. IT SHOULD BE TO, NUMBER TWO, THE FAMILY OF MAN, AND NUMBER ONE, THE PLANET AT LARGE. THIS IS OUR HOME, AND THIS IS ALL WE'VE GOT.'

– SCOTT CARPENTER, MERCURY 7

SYMM

ETRY

The Universe in a snowflake

I love the photograph of Wilson 'Snowflake' Bentley (see plate section, page two); a tilt of the head, content, protected from the cold by curiosity, absorbed in Nature's detail which he holds carefully in both hands, oblivious to the snow falling on his hat. No gloves. As a 15-year-old farm boy from Jericho, Vermont, Bentley spent the snow days from November to April with a battered microscope sketching snowflakes before they melted away. Frustrated by their transience, too short-lived to capture in detail, he began experimenting with a camera and, on 15 January 1885, he took the first ever photograph of a snowflake. Over the next 45 years he collected over 5000 images and dedicated his life to carefully observing and documenting the raindrops, snowfalls and mists that swept across his farm.

These delicate snapshots of a world available to everyone but rarely seen captured the public imagination. How could they not? They are magical, even today in an age familiar with photography. I challenge anyone to look at these structures, endless and most beautiful – to paraphrase Darwin – and not be curious. How do they form? What natural mechanism could mimic the work of a crazed, impatient sculptor obsessed with similarity and yet incapable of chiselling the same thing twice?

These are questions that can be asked about any naturally occurring structure, and which Darwin famously answered for living things in *On the Origin of Species*. In May 1898 Bentley co-wrote an article for *Appletons' Popular Science* with George Henry Perkins, Professor of Natural History at the University of Vermont, in which he argued that the evidence he'd collated frame by frame revealed that no two snowflakes are ever alike. 'Every crystal was a masterpiece of design and no one design was ever repeated', he wrote. Their uniqueness is part of their fascination and romance, yet there is undoubtedly something similar about them; they share a 'six-ness'. Which is more interesting? Perhaps it depends on the character of the observer.

Johannes Kepler is best known for his laws of planetary motion. He pored over the high-precision astronomical observations of the Danish astronomer Tycho Brahe, just as Snowflake Bentley pored over his photographs, and he noticed patterns in the data. These patterns led him to propose that planets move in elliptical orbits around the Sun, sweeping out equal areas in equal times and with orbital periods related to their average distances from the Sun. Kepler's empirical laws laid the foundations upon which Isaac Newton constructed his Law of Universal Gravitation, published in 1687; arguably (I would say unarguably, but one has to keep argumentative historians happy) the first modern scientific work.

In December 1610, shortly after the publication of two of his three laws in *Astronomia Nova*, Kepler was walking across the Charles Bridge in Prague through the Christmas dark when a snowflake landed on his coat. The evident structure of the elegant, white near-nothing interested him, and he wrote a small book entitled *On the Six-Cornered Snowflake*. It is a piece of scientific writing that transcends time and provides an illuminating and entertaining insight into a great mind at play. The title page of the book is addressed 'To the honorable Counselor at the Court of his Imperial Majesty, Lord Matthaus Wacker von Wackenfels, a Decorated Knight and Patron of Writers and Philosophers, my Lord and Benefactor'. Modern language lacks a certain flourish; I wish I had something equally magnificent with which to begin this book.

As a modern research proposal, Kepler's *Six-Cornered Snowflake* would fall at the first hurdle because it begins: 'I am well aware how fond you are of Nothing, not so much on account of its inexpensive price as for the charming and subtle *jeu d'esprit* of playful Passereau.[1] Thus, I can easily tell that a gift will be the more pleasing and welcome to you the closer it comes to nothing.' Now there's a statement of projected economic impact; the closer my research comes to nothing, the more valuable it is. Stick that on your spreadsheet... Kepler doesn't succeed in explaining the structure of snowflakes – how could he? A full explanation requires atomic theory and a good fraction of the machinery of modern physics; we will get to that later on. What he does achieve is to make vivid the joy of science; the idea that the playful investigation of Nature has immense value, irrespective of the outcome. His book explodes with excited curiosity, fizzing with speculations on snowflakes and their similarities to other regular shapes in the natural world; five-petalled flowers, pomegranate seeds and honeycombs. He covers so much ground, bouncing thrillingly from subject to subject, that eventually, with magnificent perspicacity, he has to rein himself in: 'But I am getting carried away foolishly, and in attempting to give a gift of almost Nothing, I almost make Nothing of it all. For from this almost Nothing, I have very nearly recreated the entire Universe, which contains everything!'

Kepler does have a clear question, however, which surely occurs to anyone who studies Snowflake Bentley's exquisite photographs: how do structures as ordered and regular as snowflakes form from apparently form-less origins? 'Since it always happens, when it begins to snow, that the first particles of snow adopt the shape of small, six-cornered stars, there must be a particular cause; for if it happened by chance, why would they always fall with six corners and not with five, or seven, as long as they are still scattered and distinct, and before they are driven into a confused mass?'

Kepler knew that snow forms from water vapour, which has no discernable structure. So how does the snowflake acquire structure? What is the 'six-ness' telling us about the building blocks of snowflakes and the forces that sculpt them? This is a modern way of looking at the world, one that any physicist would recognise. Kepler's insight, and his delighted frustration at not possessing the knowledge to approach an answer, echoes loudly down the centuries. 'I have knocked on the doors of chemistry,' he writes, 'and seeing how much remains to be said on this subject before we know the cause, I would rather hear what you think, my most ingenious man, than wear myself out with further discussion. Nothing follows. The End.'

Science is delighted frustration. It is about asking questions, to which the answers may be unavailable – now, or perhaps ever. It is about noticing regularities, asserting that these regularities must have natural explanations and searching for those explanations. The aim of this chapter, inspired by Kepler and Snowflake Bentley, is to seek explanations for the complex shapes in Nature; from beehives to icebergs; planets to free-diving grandmothers (honestly!). This will lead us to think about how such diversity and complexity can emerge from laws of Nature that are few in number and simple in form. At the end of the chapter, we will explain the structure of snowflakes.

[1] It is generally accepted that 'Passereau' is used here by Kepler as a pun, connecting a playful sparrow with French poet Jean Passerat, who wrote a New Year's poem on the subject of Nothing. Obscure, but fun!

Why do bees build hexagons?

Bees have a got a tricky problem to solve. How do you store honey, the food that will sustain your colony, through the long winter months? We know that bees build honeycombs for this purpose. Kepler was interested in the structure of honeycombs precisely because they are built, as he writes, by 'an agent'. Since he was seeking the 'agency' that sculpts snowflakes, he decided to search for the reason why bees build hexagons. With the benefit of Darwin, we might propose that the answer will involve natural selection, which is a simple and powerful idea. If an inherited trait or behaviour confers an advantage in what Darwin referred to as the 'struggle for life', that trait will come to dominate in future generations simply because it is more likely to be passed on. The sum of an organism's physical characteristics, behaviours and constructions is known as the phenotype, and it is on this that natural selection operates. If natural selection is the reason for the structure of honeycombs, we should be able to understand why their hexagonal shape offers an advantage to the bees that construct them.

Charles Darwin was fascinated by bees and followed precisely this path. 'He must be a dull man who can examine the exquisite structure of a comb, so beautifully adapted to its end, without enthusiastic admiration', he wrote in *On the Origin of Species*. I enjoy the directness of Victorian writing; if your mind isn't inquisitive, you are a dullard. In the same seminal work, Darwin describes a series of experiments he conducted in order to understand the cell-making instincts of the hive bee.

'... it seems at first quite inconceivable how they can make all the necessary angles and planes, or even perceive when they are correctly made. But the difficulty is not nearly so great as it first appears: all this beautiful work can be shown, I think, to follow from a few very simple instincts.'

To identify these simple instincts, Darwin compared the hive-making behaviours of the honeybee with a less architecturally accomplished species of bee, the Mexican *Melipona domestica*. The Melipona bees construct regular combs of cylindrical cells which Darwin asserted to be a simpler geometrical form, intermediate between no structure at all and the hexagons of the honeybees. 'We may safely conclude that if we could slightly modify the instincts already possessed by the Melipona, this bee would make a structure as wonderfully perfect as that of the hive bee.'

To test the hypothesis, Darwin conducted a series of experiments in collaboration with his friend and fellow naturalist William Bernhardt Tegetmeier. They added different-coloured dyes to the beeswax, enabling them to create a visual record of the construction process, and were able to conclude that the bees first build cylindrical cells that are subsequently modified to form hexagons. Darwin was able to describe this in terms of natural selection:

'Thus, as I believe, the most wonderful of all known instincts, that of the hive-bee, can be explained by natural selection having taken advantage of numerous, successive, slight modifications of simpler instincts; natural selection having by slow degrees, more and more perfectly, led the bees to sweep equal spheres at a given distance from each other in a double layer, and to build up and excavate the wax along the planes of intersection. The bees, of course, no more knowing that they swept their spheres at one particular distance from each other, than they know what are the several angles of the hexagonal prisms and of the basal rhombic plates. The motive power of the process of natural selection having been economy of wax; that individual swarm which wasted least honey in the secretion of wax, having succeeded best, and having transmitted by inheritance its newly acquired economical instinct to new swarms, which in their turn will have had the best chance of succeeding in the struggle for existence.'

Darwin concluded that bees build hexagonal honeycombs because they are the most economical way of dividing up their honey storage area. Hexagons use less wax, and the bees that use less wax are more efficient and more likely to survive and pass on their inherited behaviour to the next generation. This makes sense, because building a wax hive is extremely honey-intensive; for every gram of wax a bee produces it has to consume up to eight grams of honey. There is clearly an impetus to build efficiently, since using as little wax as possible maximises the honey available for food – an advantage that will have shaped the behaviour of honeybees over generations.

Is this correct? It's certainly plausible. If bees used cylinders to build their honeycomb there would be gaps between each cell and the whole structure would be less efficient. Similarly, pentagons and octagons also produce gaps and so cannot be optimal. It is possible to imagine that each cell could be constructed in a bespoke shape by each bee to fit perfectly with its neighbour. In this 'custom-made' scenario each cell would be a different shape, but the gaps in the honeycomb could still be minimised. A problem with this strategy might be that one bee has to finish before the next bee can create a cell to fit. That's an inefficient use of time. A repeatable single shape that leaves no gaps would seem to be preferred. The square, the triangle and the hexagon are the only regular geometrical figures that can fit together in a plane without leaving gaps.[2]

But why do bees use hexagons? Sometime around 36 BC, the Roman scholar Marcus Terentius Varro wrote down the earliest-known description of the honeycomb conjecture. This states that the most efficient way to divide a surface into regions of equal area (cells) with the least total perimeter (wax) is to use a regular hexagonal grid or honeycomb. No proof was offered, and the assertion remained conjecture for the next 2000 years until, in 1999, a mathematician at the University of Michigan named Thomas Hales found a proof: a hexagonal pattern is the most efficient engineering design. Natural selection, selecting for efficiency and creating structures that are a shadow of an elegant underlying mathematical law. What a beautiful answer to a simple question.

‘BEES, THEN, KNOW JUST THIS FACT WHICH IS USEFUL TO THEM – THAT THE HEXAGON IS GREATER THAN THE SQUARE AND THE TRIANGLE AND WILL HOLD MORE HONEY FOR THE SAME EXPENDITURE OF MATERIAL IN CONSTRUCTING EACH.’

— PAPAS OF ALEXANDRIA, AD 340

Well ... possibly, but there may be more to it. In 2013, three engineers – Karihaloo, Zhang and Wang – published an article entitled 'Honeybee combs: how the circular cells transform into rounded hexagons'. The claim is that honeybees, just like the Melipona bees that Darwin dismissed as inferior architects, make cells that are initially circular in cross section. The hexagons appear because the bees' body heat softens the wax until it reaches 45 degrees Celsius, a temperature at which wax begins to flow like a viscous fluid. The circular cells of molten wax then act in a similar way to soap bubbles, joining together at an angle of 120 degrees wherever they meet. If all the bubbles or wax cells are identical in size and spacing, the circular cells spontaneously reform into a sheet of hexagons. Karihaloo and his team demonstrated this by using smoke to interrupt honeybees in the process of making a hive, revealing that the most recently built cells were circular, whilst the older ones had developed into hexagons. This transition from cylindrical to hexagonal structure appears to be what Darwin observed, but the explanation for the transition is different.

Natural selection is still the basic explanation for the hexagons, but the bees don't have to go to the trouble of building the most efficient packing shape because physics will do that for them, given a nice sheet of circular cells of similar size and spacing and some body heat. To me, this is even more elegant and efficient; the bees allow physics to finish their work! As the authors of the study write: 'We cannot ... ignore, nor can we not marvel at the role played by the bees in this process by heating, kneading and thinning the wax exactly where needed.' Is this the solution to a problem that has fired the imagination of so many for so long? The origin of the hexagons continues to be debated, and Karihaloo *et al.* will probably not be the last word in the literature.

This is as it should be, and illustrative of something that is often missed in the presentation of science. Scientific results are always preliminary. No good scientist will believe that they have offered the last word on a given subject. A result is published if the authors and a group of their peers consider it to be a valuable contribution to the field. Crucially, this does not mean that it's correct; it means that it's not obviously wrong. Rather than closing down a question, publication is intended to be a red flag to bullish colleagues. As one reads in Kepler's partial, yet evident, delight in not discovering a satisfactory explanation for the structure of a snowflake, there is joy in hearing what you think, my most ingenious colleague.

[2] There are many ways of tiling a plane using tiles of more than one shape; Penrose tiling is a particularly interesting example, in which the plane can be tiled by a set of 'aperiodic' tiles that form a pattern that never repeats.

Knocking on the doors of chemistry

In the final lines of *The Six-Cornered Snowflake*, Kepler writes with lovely regret that he is 'knocking on the doors of chemistry'; the implication being that those doors would be opened by future generations. He asserts, correctly, that the structure of the snowflakes must be due at least in part to some underlying structure or shape, but given that atomic theory didn't move into the realm of experimentally testable science until the early nineteenth century, and the structure of atoms themselves was a twentieth-century discovery, Kepler had no way of unlocking the doors. We now know that the building blocks of snowflakes are water molecules, and water molecules are capable of extremely complex behaviour when they get together. That may be a surprising statement if we think of water as the colourless, odourless liquid in a glass. Perhaps it shouldn't be so surprising if we think of water molecules as the objects that come together spontaneously to produce the romantic flourishes of form and exquisite diversity of snowflakes.

Single water molecules aren't particularly complicated. They are molecules of hydrogen and oxygen, bonded together. Oxygen was first isolated in 1774 by Joseph Priestley, the son of a Yorkshire woollen cloth maker, and Henry Cavendish first identified hydrogen in 1766. The Nobel Prize in Physics in 1926 was awarded to Jean Baptiste Perrin for the confirmation of the physical reality of molecules, just about within living memory, which demonstrates how difficult it is to study the microscopic world and how quickly cutting-edge science can become common knowledge.

A water molecule consists of two hydrogen atoms bonded to a single oxygen atom: H_2O (see illustration opposite). The water molecule isn't linear – the hydrogen atoms are displaced at an angle of 104.5 degrees. The reason for this is the presence of two extra pairs of electrons that sit on the opposite side of the oxygen atom. To see why that is, let's have a very brief tutorial on atomic physics and quantum mechanics.

Atoms are made up of three constituents, as far as chemists are concerned (we'll dig more deeply into this later on); they consist of a small, dense, atomic nucleus made up of protons and neutrons, with electrons orbiting a long way away. If the nucleus were the size of a tennis ball, the outer electron orbits would be several kilometres across. Hydrogen is the simplest element; its nucleus consists of a single proton. Next is helium, which contains two protons and two neutrons. Oxygen has eight protons and eight neutrons. The nucleus is surrounded by electrons, which are held in place by one of the four fundamental forces of Nature: electromagnetism. Electrons are negatively charged and protons are positively charged, and the negative electric charge of the electron is precisely equal in magnitude but opposite in sign to the positive electric charge of the proton. Nobody knows why these charges are precisely equal in magnitude; it's one of the great mysteries of fundamental physics. The atoms of each chemical element are electrically neutral, which means that the number of protons in the nucleus is equal to the number of electrons that surround it. Hydrogen atoms have a single electron, therefore, whilst oxygen atoms have eight electrons.

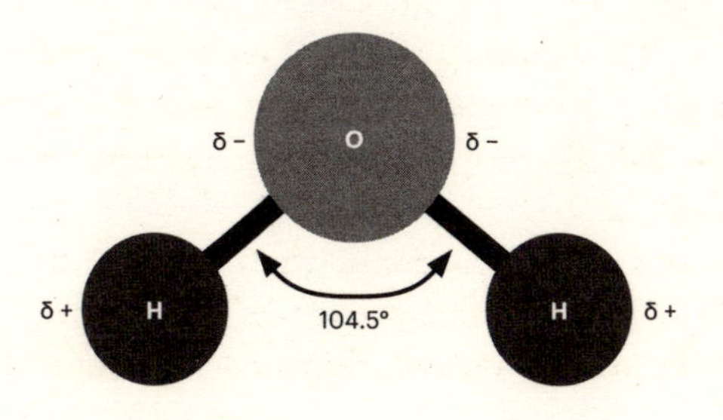

Below: The structure of a water molecule, showing oxygen's eight electrons, two of which are shared with the hydrogen atoms.

Bottom: The hexagonal crystalline structure of Ice 1_h. Water molecules are attached together by hydrogen bonds, with oxygen atoms from one water molecule lining up with hydrogen atoms from another.

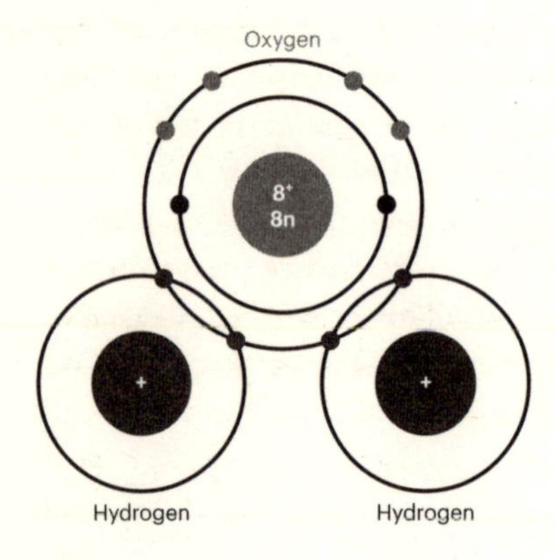

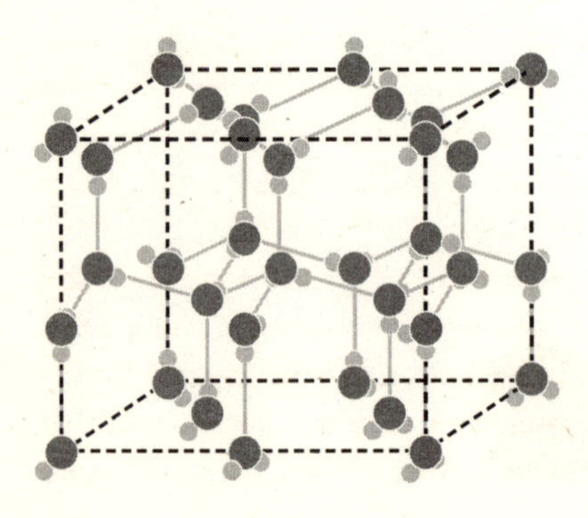

Now we need a little sprinkle of quantum theory. You can picture the electric charge of the atomic nucleus as creating a kind of box within which the electrons are trapped. Electrons, along with all of the fundamental building blocks of the Universe, obey the laws of quantum theory, which describe how they move. It turns out that the basic rules of quantum theory are counterintuitive and fly in the face of common sense. But that is okay because there is no reason at all to expect the laws that govern the Universe to be in accord with 'common sense'. The most fundamental rule governing the behaviour of subatomic particles is that they don't like to stand still. Unfettered, they are very likely to wander off, and the more we try to pin them down, the more they are inclined to wander. The presence of the nucleus tames the anarchic electrons somewhat, by confining them to the 'nuclear box'.

Another rule governing the behaviour of electrons is that they don't much like each other's company. This is known as the Pauli exclusion principle, also a consequence of the laws of quantum theory. Electrons will arrange themselves around the nucleus such that they stay away from each other, as best they can. There is a caveat, though, which is important for understanding the structure of atoms. Electrons of opposite spin are allowed to get close together (or 'pair up'). Of course they cannot get too close because they have the same electric charge and 'like-charges repel'. Spin is a property of subatomic particles that is easy to name but hard to picture. You could think of electrons as little spinning tops, if you like, but that's a bad analogy on many levels, so you probably shouldn't. Having said that, spin is a measure of how much an electron is spinning – it is just that the notion of a spinning point is not something we can easily imagine. For particles such as electrons, which are known as 'spin ½' particles or fermions, spin can have only two values; these are known as spin-up and spin-down. Spin is a direct, if rather subtle, consequence of the merger between Einstein's Theory of Special Relativity and quantum theory, achieved by physicist Paul Dirac in his equation describing the electron in 1928. The details don't matter here; what matters is that the negatively charged electrons get trapped by the positive electric charge of the protons in the atomic nucleus and that electrons tend to keep away from each other, although opposite-spin

ANOTHER RULE GOVERNING THE BEHAVIOUR OF ELECTRONS IS THAT THEY DON'T MUCH LIKE EACH OTHER'S COMPANY. THIS IS KNOWN AS THE PAULI EXCLUSION PRINCIPLE.

electrons can get closer together than same-spin electrons can. This is enough information for us to get a basic understanding of a water molecule. Oxygen has eight electrons. Two of the electrons sit close to the nucleus and do not play much of a role in binding the two hydrogen atoms to the oxygen. The remaining six are shared out as in the diagram on page 26.[3]

One of the basic concepts in chemistry, which again goes all the way back to the fundamental laws of quantum theory, is that electrons can be shared between atoms. This results in the formation of a chemical bond. Two hydrogen atoms will share their single electrons with an oxygen atom if they can, pairing up to fill the two remaining outer slots around the oxygen nucleus; the result is a water molecule, which is shown in the top illustration. The reason for the 104.5-degree 'kink' is the presence of the other two pairs of electrons in the outer level of the oxygen atom. They take up residence on the opposite side of the oxygen atom to the hydrogen atoms, giving the water molecule its distinctive shape, and its many unusual properties.

The water molecule, like its constituent atoms, is electrically neutral, but the uneven distribution of electrons means that the hydrogen atom 'legs' have a very small net positive charge, whilst the oxygen end of things has a slight net negative charge. Water is known as a polar molecule for this reason – it has a negative end and a positive end. This opens up a world of complexity.

An important consequence of water's polarity is that water molecules like to stick together. The negatively charged oxygen ends of water molecules attract the positively charged hydrogen ends of other water molecules and they attach together through what is known as a hydrogen bond. This happens to an extent in liquid water, resulting in quite large and complex structures.

The effects are even more dramatic when temperatures drop and water freezes to form ice. Water ice is very weird stuff. There are seventeen known forms of ice, the most common of which on Earth is called Ice 1_h (the structure of which is shown in the lower illustration). The regular crystalline structure leads to one of water's most bizarre properties: ice floats. This is very unusual behaviour. Every other commonly occurring solid is denser in the solid phase than in the liquid phase, and therefore does not float on its own liquid. The crystalline structure of ice, however, is so open that at atmospheric pressure and 0 degrees Celsius it is 8 per cent less dense than liquid water. This is why icebergs float on the oceans.

This is interesting, and it isn't necessarily a trivial observation. It has been suggested that this unusual behaviour may have played a vital role in the evolution and persistence of life on Earth. If ice were denser than liquid water, sea ice would sink to the ocean floor. In such a scenario, particularly during Earth's great glaciations, the lakes, seas and oceans of Earth could have frozen from the bottom up, perhaps becoming permanently solid. This would have had a dramatic impact on the ecosystems and food webs that rely on the bottom-dependent animal and plant life in fresh and seawater.

The complex structure of ice is a consequence of the laws of quantum theory, which are small in number and simple. By simple, we don't mean to suggest that quantum theory is a simple thing to learn and apply; it isn't. The mathematics can be technically difficult. Quantum theory is simple in the sense that it consists of a small number of mathematical rules that describe a wide range of natural phenomena of all sizes, from the structure of atoms and molecules to the nuclear reactions in the Sun. They also describe the action of real-world devices such as transistors and lasers and, more recently, exotic pieces of technology such as quantum computers.

A tremendous economy of description is one of the defining and most surprising features of modern science; it is not *a priori* obvious that a small collection of fundamental laws should be capable of describing the limitless complexity of objects that populate our Universe, and yet this is what we have discovered over the last few centuries. Perhaps a universe regular enough to permit the existence of natural objects as complex as the human brain must be governed by a simple set of laws, but since we do not yet understand the origin of the laws, we do not know. It is interesting that such complexity can emerge from underlying simplicity, however, and the humble water molecule is a good example. Its asymmetrical 'kinked' structure, which is ultimately responsible for the complex structure of ice, is a consequence of the laws of quantum theory, but these laws do not have 'kinks' built into them. Indeed, a physicist would say that the laws are possessed of a high degree of symmetry, as are the nuclei of hydrogen and oxygen; they form nicely spherical 'nuclear boxes' to trap the electrons. But bring them together and they form an asymmetrical structure.

The concept of symmetry is central to modern physics, and we'll meet it throughout this book. For now, let us simply note that the asymmetric structure of the water molecule is a consequence of the way that electrons fit around the nucleus of an oxygen atom. It is because there are four available outer slots and six electrons to fill them that an asymmetric molecular structure results when two hydrogen atoms approach the oxygen, and that structure emerges spontaneously. Nobody had to design the water molecule and make an aesthetic choice about the 104.5-degree bond angle! It's a consequence of, but not arbitrarily inserted into, the laws of quantum theory.

The properties of water are ultimately a result of the interactions between molecular building blocks. In turn, the properties of water molecules are a result of the interactions between their constituents – hydrogen and oxygen atoms. The properties of hydrogen and oxygen atoms are a result of the interactions between their constituents – protons, neutrons and electrons – and these interactions are governed by a simple set of rules. Is this infinite regression? How far can we go, digging deeper and deeper for more fundamental explanations for the properties of matter in general?

[3] The details of where the electrons reside in any particular molecule are generally very hard to compute. The details follow from solving the Schrödinger equation in the spherically symmetric potential generated by the atomic nucleus. If you'd like more details, and are interested in delving more deeply into quantum theory, there is much more in my book with my colleague Jeff Forshaw entitled *The Quantum Universe*.

The fundamental building blocks and the forces of Nature

It was twenty years ago today that I began my PhD. Today is 1 October 2015. Three years later I submitted my thesis 'Double Diffraction Dissociation at Large Momentum Transfer'. I was interested in the behaviour of an object known as the Pomeron, named after the Russian physicist Isaak Pomeranchuk. I looked for it in the debris of high-energy collisions between electrons and protons, generated by a particle accelerator known as HERA. HERA is the wife of Zeus, and also the Hadron-Electron Ring Accelerator. The machine was 6.7 kilometres in circumference, located below the streets of northern Hamburg, which is a beautiful city in which to be a student. In the winter, the River Elbe freezes, but icebreakers clear a path to the port and the city feels proximate to the Baltic. In summer the small beaches that line the river beneath the old houses of Blankenese are busy and the city feels Mediterranean. In the early mornings at any time of year, a deracinated twenty-something from Oldham can be distracted on the Reeperbahn. It's a remarkable thing that someone can spend three years looking at the fine detail of high-energy collisions between electrons and protons, hunting for a thing called a Pomeron.

Why was I interested in Pomerons? I was engaged in testing our best theory of one of the four fundamental forces of Nature. We've met one of these forces already – electromagnetism – which holds electrons in orbit around the atomic nucleus and water molecules together via hydrogen bonds. My investigations of the Pomeron were concerned with exploring another of the four – the strong nuclear force. The need for such a force is clear if you think about our description of the oxygen nucleus. It is a tightly knit ball of eight positively charged protons and eight uncharged neutrons. One of the fundamental properties of the electromagnetic force is that like-electrical charges repel each other; in which case, why doesn't the atomic nucleus blow itself apart? The answer is that the strong nuclear force sticks the nucleus together, and it is far stronger than the electromagnetic repulsion between the protons.

Protons are small, but they make up just over half of you by mass. Most of the rest of you is made of neutrons. There are around twenty thousand million million million million protons in the average human being. In scientific notation, that's 2×10^{28}, which means 2 followed by 28 zeros. You are pretty simple at this level.

When you look deeper into the heart of the protons and neutrons themselves, things appear to get more complicated. Protons are small by everyday standards, but it is well within our current scientific and engineering capabilities to measure their size and look inside them. This is what HERA was designed to do. The machine was a giant electron microscope, peering deep into the heart of matter. You have to define what is meant by size carefully, because a proton doesn't have a hard edge to it, but recent measurements put its radius at just over 0.8 femtometres, which is just under 10^{-15} m – a thousand million millionths of a metre.[4]

Below: The neutral current DIS process via photon exchange.

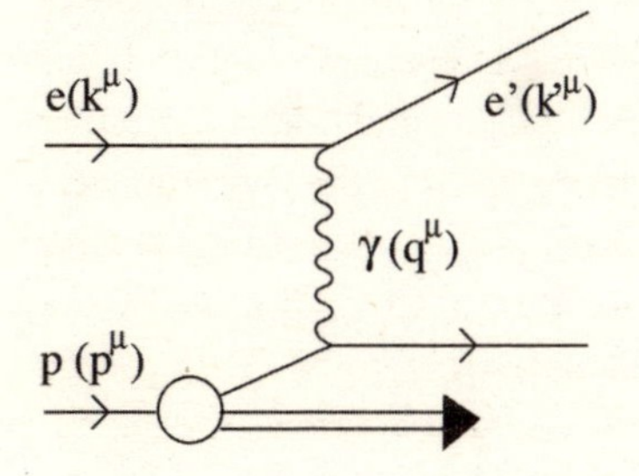

This spread: $F_2(x, Q^2)$ as measured at HERA, and in fixed target experiments, as a function of Q^2 (a) and x (b). The curves are a phenomenological fit performed by H1 [26]. $c(x)$ is an arbitrary vertical displacement added to each point in (a) for visual clarity, where $c(x) = 0.6(n - 0.4)$, n is the x bin number such that $n = 1$ for $x = 0.13$.

A

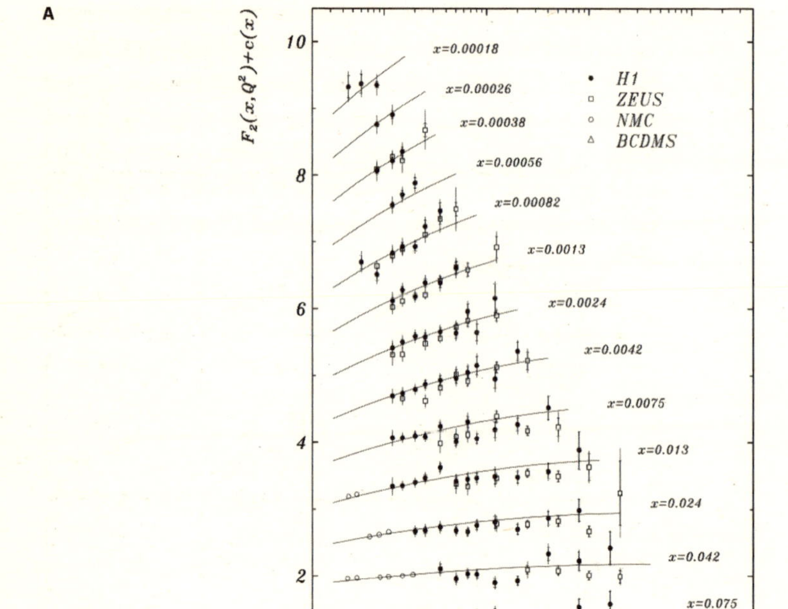

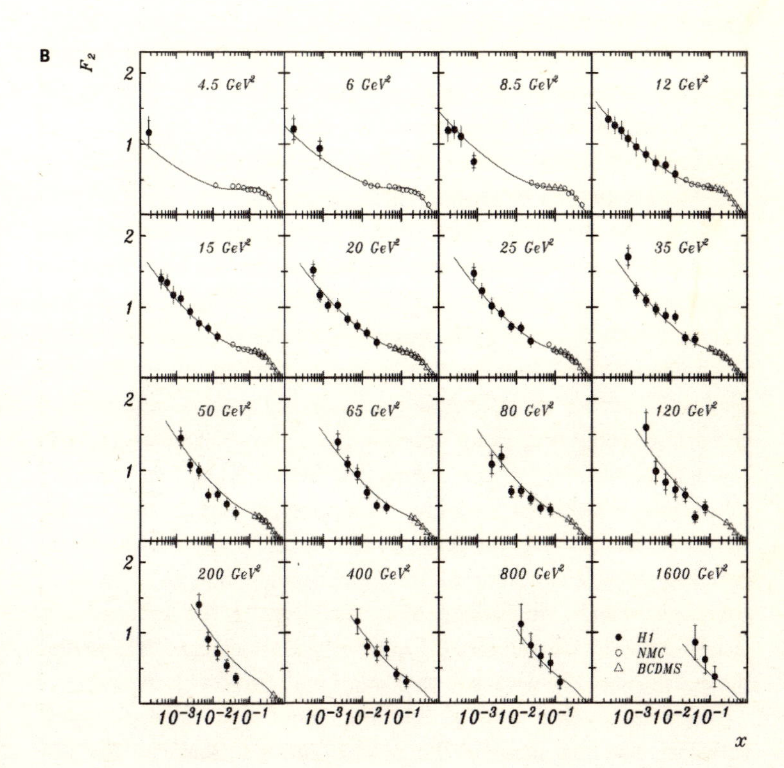
B
F_2
4.5 GeV²
6 GeV²
8.5 GeV²
12 GeV²
15 GeV²
20 GeV²
25 GeV²
35 GeV²
50 GeV²
65 GeV²
80 GeV²
120 GeV²
200 GeV²
400 GeV²
800 GeV²
1600 GeV²
H1
NMC
BCDMS
10^{-3} 10^{-2} 10^{-1}
x

Because I'm getting old and sentimental, but also in service of the narrative, I've indulged myself and included two plots from the thesis I wrote in Hamburg twenty years ago. After all, this was my snowflake. The first one shows a drawing I made using a 1990s UNIX computer program called *xfig* (see illustration, previous page). Happy days. It shows an electron colliding with a proton. The language of modern physics is superficially opaque, as evidenced by the caption of my thesis figure, but the language isn't designed to make physicists appear clever. To be honest, I never thought a non-physicist would read it. Every word is necessary and means something. George Orwell would approve. 'A man may take to drink because he feels himself to be a failure, and then fail all the more completely because he drinks. It is rather the same thing that is happening to the English language. It becomes ugly and inaccurate because our thoughts are foolish, but the slovenliness of our language makes it easier for us to have foolish thoughts.'[5] Physics is about precision of thought, which is aided and evidenced by precision of language.

Here is the meaning of the caption. Neutral current means that the electron bounces off the proton by exchanging an electrically neutral object with it – in this case, a photon; a particle of light. The photon is shown in the diagram as the wavy line, labelled by the Greek letter γ. DIS stands for 'Deep Inelastic Scattering', which means that the photon is hitting something deep inside the proton, resulting in the proton being broken into pieces. This is how a modern particle physicist would describe the interaction between any two particles; interactions involve the 'exchange' of some other particle that carries the force. In this case, the force is electromagnetism and the force-carrying particle is a photon. The most fundamental description of the mechanism by which water molecules stick together to form ice is that photons are being emitted and absorbed by electrons in the water molecules, with the net result that water molecules stick together.

There is another way of thinking about this electron–proton collision. You can imagine the photon emitted from the electron smashing into the proton and revealing its inner structure. That structure is shown in the second figure from my thesis, shown opposite.

Allow me a single paragraph of postgraduate-level physics. I want to take this liberty for two reasons. The first is that there is great joy to be had in understanding a complex idea, and in doing so glimpsing the underlying simplicity and beauty of Nature. The biologist Edward O. Wilson coined the term 'Ionian Enchantment' for this feeling, named after Thales of Miletus, credited by Aristotle as laying the foundations for the physical sciences in 600 BC on the Greek island of Ionia. The feeling is one of elation when something about Nature is understood, and seen to be elegant. The second reason is to revisit and enhance an idea we've been developing. Science is all about making careful observations and trying to explain what you see. That might be the hexagonal structure of a beehive, the jagged symmetry of a snowflake, or the details of how electrons bounce off protons. Careful observations lead to Ionian Enchantment.

At HERA, we measured the angle and energy of the electrons after they hit the protons. This is a simple thing to do, and it allowed us to build up a picture of what the electron 'bounced off' – the fizzing heart of matter. Two different ways of visualising the inside of a proton are shown in the figure. The thing called F_2 (x,Q^2) is known as the proton structure function. Now for the precise bit of observation that requires thought. Have a look at illustration (a) on page 34 and focus on the bottom line of the graph labelled $x = 0.13$. The points along this line tell you the probability that an electron will bounce off something inside the proton that is carrying 13 per cent of the proton's momentum – this is what $x = 0.13$ means. The quantity Q^2 is known as the virtuality of the photon that smashes into the proton. One way to think about this quantity is as the resolving power of the photon. High Q^2 corresponds to short wavelength, which means that high Q^2 photons can see smaller details. The $x = 0.13$ line is pretty flat, which means that whatever the photon is bouncing off, it behaves as if it has no discernable size. This is because what we see does not change as we crank up the resolving power of the microscope (which corresponds to going to higher Q^2), and this is what would happen if the photon were scattering off tiny dots of matter inside the proton. The dot is known as a quark, and as far as we can tell, it is one of the fundamental building blocks of the Universe. Together, these two plots describe in detail the innards of the proton as revealed by years of experimental study by many hundreds of scientists at the HERA accelerator.

The proton is a seething, shifting mass of dot-like constituents, continually evolving around scaffolding. The scaffolding consists of three quarks; two 'up' quarks and one 'down' quark. The quarks are bound together by the strong nuclear force, which is carried by particles called gluons in much the same way that the electromagnetic force is carried by photons. Unlike photons, however, the gluons can interact with each other through the exchange of more gluons, and that results in the proton having an increasingly complex structure as we dial up the resolving power. Illustration (b) shows this behaviour; the rising curves towards smaller x are telling us that there is a proliferation of gluons, each carrying very small fractions of the proton's momentum. Illustration (a) also shows this. The lines are not flat at smaller x. In the jargon, this behaviour is known as 'scaling violation', which means that as we dial up the resolving power the dot-like constituents appear to be increasingly numerous. In other words, at low resolving power we tend to resolve only the scaffolding, i.e. the three quarks, while at high resolving power the full glory of the proton's gluonic structure is revealed to us. Roughly speaking, gluons carry around half of the momentum of a proton, because there are so many of them buzzing around between the quarks. The lines on these graphs, which go pretty much through the data points, are calculated using our best theory of the strong nuclear force: Quantum Chromodynamics, or QCD. QCD is a set of rules that specifies the probability that a quark will emit a gluon, and also how gluons interact with other quarks and gluons. It's a quantum theory – the same basic framework we referred to when we discussed the structure of the water molecule. When we are dealing with electric charges – for example, the interactions between electrons and the atomic nucleus – we use our quantum theory of electromagnetism called Quantum Electrodynamics, or QED.

I remember writing computer programs to skim through vast amounts of data about individual electron-proton collisions and make figures like the one above. On the computers we had in the 1990s these programs took days to run. Even now, looking at these plots, I find it exhilarating to consider that I'm looking at the structure of an object a thousand million millionths of a metre in size, measured using a machine 6.7 kilometres in circumference beneath the city of Hamburg, and that we have a theory that allows us to understand and describe what we see. Industrial engineering and subatomic beauty in concert. The Ionian Enchantment.

On the next page you will find a snapshot of the deep structure of ordinary matter. You are this, at the level of accuracy we can measure today. Two sorts of quarks, stuck together by gluons, to make protons and neutrons that are stuck together by more gluons to make atomic nuclei. Electrons are stuck in orbit around the nuclei by photons to make atoms and atoms stick together by exchanging photons between their electrons to make molecules. And so it goes! This simple picture is the result of a hundred years of experimental and theoretical investigation. The structure of everything can be explained using a set of building blocks and some rules. We've met three of the building blocks; up quarks, down quarks and electrons. We've also met two forces; the strong nuclear force and the electromagnetic force. There is another force called the weak nuclear force that can convert up quarks into down quarks, with the simultaneous emission of another sort of particle called the electron-neutrino. In total that makes four matter particles. The weak force is carried by particles known as the W and Z bosons. There is also the Higgs boson, discovered in 2012 at the Large Hadron Collider (LHC) at CERN, in Geneva, which gives the building blocks their mass.

The fourth and final fundamental force is the most familiar – gravity. It is so weak that its effects on the subatomic world are invisible even in our most high-precision experiments, like those at HERA. If this statement seems a little mystifying, particularly if you've ever fallen off a ladder, then park it in your memory for a while; we'll get back to gravity later when we discuss the shape of planets and galaxies.

These four particles, four forces and the Higgs boson appear to be all that is needed to make a water molecule, a honeybee, a human being, or planet Earth. This is a dazzlingly elegant and simple structure. For some reason, Nature didn't adopt this economical scheme but instead made two further copies of the family of up quarks, down quarks, electrons and electron neutrinos. These two extra families are identical to the first family in every way except that they are more massive, possibly because they interact with Higgs particles in a different way. The existence of the three families of particle is another of the great mysteries, and discovering why Nature appears to have been unduly profligate is one of the most important goals of twenty-first-century particle physics. She won't have been unduly profligate, of course! We know that three families is the minimum number to accommodate a process known as CP violation, which is needed to explain why, if the Universe started out with equal amounts of matter and anti-matter, there is matter left over in the Universe today to make stars and people. But that's not an answer to the 'Why?' question, and it would be nice to know if the existence of planets, stars and galaxies is down to more than blind luck.

With these extra families, there are twelve fundamental particles of matter, four different sorts of force-carrying particle and the Higgs particle. That's it, as far as we know – although I wouldn't be surprised if some more pop up at the Large Hadron Collider over the next few years. This is fuelled by the fact that we already have good evidence from many independent astronomical observations that there is another form of matter in the Universe known as dark matter. There is five times more dark matter than 'normal' matter in the Universe by mass, and the dark matter cannot be made up out of the twelve particles that we've seen in experiments at particle accelerators such as HERA or the LHC. The collection of fundamental building blocks, circa 2015, is shown in the illustration below.

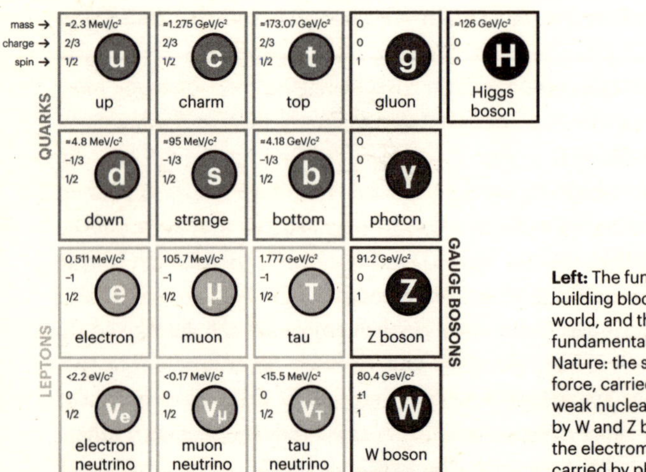

Left: The fundamental building blocks of the natural world, and three of the four fundamental forces of Nature: the strong nuclear force, carried by gluons; the weak nuclear force, carried by W and Z bosons; and the electromagnetic force, carried by photons.

This isn't intended to be a complete course on particle physics, much as I'd like to deliver that; rather, it is a chapter about shapes and patterns in Nature and what they reveal about the way in which the Universe works. Having said that, if you'll allow me one last foray into particle physics, the story of the discovery of the quarks inside the proton and neutron is a very beautiful example of the way physicists notice patterns and attempt to explain them. The remarkable thing is that quarks were predicted *before* they were discovered experimentally.

The theoretical prediction that building blocks exist beneath the level of protons and neutrons was made by Murray Gell-Mann and George Zweig in 1964. It was based on a pattern in the subatomic particles known at the time. By the early 1960s, an inelegant, profligate and seemingly ever-expanding list of subatomic building blocks had been discovered. The proton and neutron are part of a whole family of particles known as baryons; there are Lambdas, Sigmas, Deltas, Cascades and a host of others. There is also a family of particles known as mesons: Pions, Kaons, Rho and so on. There are thirteen different types of Lambda particle alone, nine Sigmas and eight Kaons. Particle physics was looking increasingly like a subatomic branch of botany. Then Gell-Mann and Zweig noticed a beautiful pattern. The particles could be arranged according to their observed properties in geometrical patterns. One such pattern is shown in the illustration on the next page. Today, these are known as 'super-multiplets'.

As Kepler suspected when he considered the six-fold symmetry of snowflakes, patterns in Nature are often a clue that there is a deeper underlying structure. The patterns may or may not be easy to recognise – Gell-Mann received the Nobel Prize in Physics in 1969 for noticing the pattern amongst the particles – but they are the Rosetta Stone that allows Nature's language to be deciphered. In this case, the pattern in the particles suggested to Gell-Mann and Zweig that the baryons are all constructed out of three smaller building blocks, that Gell-Mann called quarks. When they first recognised the pattern, they included three quarks in their scheme: up, down and strange. The different baryons on the lower planes of the super-multiplets are the possible three-fold combinations of the three

Below: A baryon 'super-multiplet' showing the quark content of each baryon.

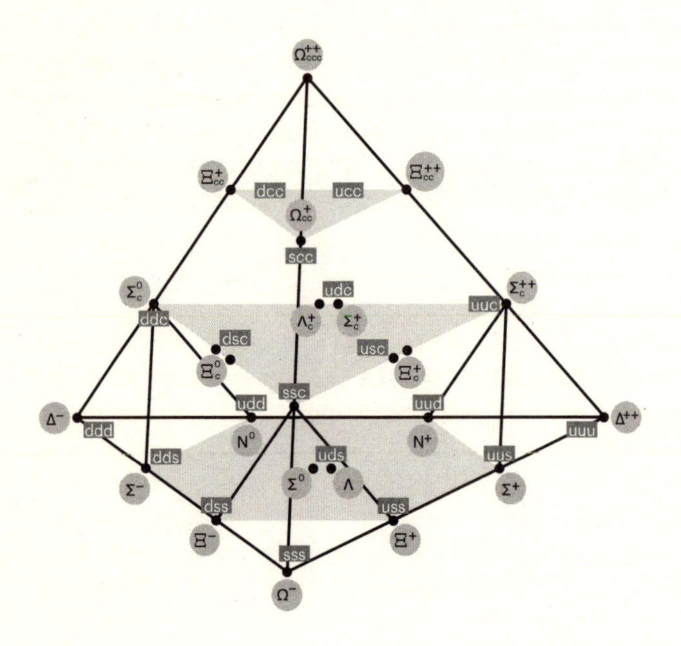

building blocks. Adding a fourth quark – charm – constructs the higher layers. The quark constituents of the particles are shown in the illustration opposite: for example the Δ^{++} contains three up quarks.

The particle on the base of the pyramid in the illustration, known as the Omega-minus, is of particular historical interest because its existence was predicted by Gell-Mann at a meeting at CERN in 1962, based solely on the pattern of the base of the pyramid. It was subsequently discovered at the Brookhaven National Laboratory in the United States in 1964. When a theory predicts the existence of something new that is subsequently discovered, we can have particular confidence that we are on the right track.

We've met three of the four fundamental forces of Nature; the strong and weak nuclear forces and electromagnetism, and the twelve building blocks of Nature. We will now turn to the final, weakest and most familiar force – gravity – and investigate it by thinking about the size and shape of the objects it sculpts. These are not tiny things like subatomic particles, or small things like snowflakes, but very much larger structures: planets, stars and galaxies.

[4] There is currently some discrepancy between different measurements of the proton radius, which may signal something interesting that we don't understand. See, for example, http://arxiv.org/pdf/1502.05314.pdf, which is a technical paper, but I recommend a glance because it demonstrates rather beautifully the precision of modern particle physics.

[5] George Orwell, *Politics and the English Language*.

Why is the Earth a sphere?

There is a photograph of our planet known as the Blue Marble. It was taken on 7 December 1972 by the crew of Apollo 17 during their journey to the Moon. Close to the winter solstice, Antarctica is a continent of permanent light, and Madagascar, the island of lemurs, takes centre stage. Ochre deserts set against blue oceans, green hues hinting at life.

On 5 December 2012, NASA released the Black Marble, an image of the Americas at night. Now we see a civilisation on the planet; the lights herald the dawn of the Anthropocene – the age of human dominance. What do we see in these images? What is the most basic property of Earth? Alexei Leonov, on completing the first human spacewalk on 18 March 1965, had an answer.

'I never knew what the word round meant until I saw Earth from space.'

Alexei Leonov, Voskhod 2, Soyuz 19/ASTP

Seen from space, the Earth is a near-perfect sphere. All the planets in the Solar System, all the large moons and the Sun itself share this property, as does every star in the Universe. Why? If lots of different objects share a common feature, there must be an explanation. To make progress, let's think about what could affect the shape of a planet, moon or star. It can't be much to do with their composition because planets are made of different stuff to stars. The Earth is made up of heavy chemical elements such as iron, oxygen, silicon and carbon. The Sun, on the other hand, is primarily hydrogen and helium; it's a giant ball of plasma with no solid surface. Giant planets such as Jupiter have more in common with stars than with Earth, at least in terms of their composition. They too are primarily composed of hydrogen and helium. Stars and planets are united, however, by the force that formed them and holds them together – gravity. So to understand why they are all spherical, we should explore the nature of the gravitational force further.

Defying gravity

For most of the time Tarragona is a quiet Mediterranean port on the northeastern coast of Spain, but each September it explodes into vivid, violent colour as teams compete against gravity in the Tarragona *Castells* competition. *Castells* are human towers, reaching ten people high and involving an intricate mix of strength, balance, strategy and teamwork to be built up to the top. Each team begins by forming the foundations of the tower, with up to two hundred people creating the *pinya*. Once the foundation is in place, a variety of human geometries are used to build as high as possible, with each level taking shape before the next is added. The most successful team is the Castellers de Vilafranca, having won the Tarragona competition eight times since 1972. A mass of green shirts acting in unison flows from one level to the next, with higher levels consisting of fewer people, until two children form a final stable platform for the *enxaneta* – the *casteller* who ascends daringly to the summit; since low mass, agility – and perhaps a lack of fear – are called for, the *enxaneta* will be as young as 6 or 7 years old. This is what the crowds have come to see. Towers give way, human buildings come tumbling down, falls softened by the elbows, knees, heads and shoulders, colliding and crashing, usually delivering only bruises, bumps and the occasional lost tooth. Serious injuries are very rare.

It is obvious why people fall to the ground if they lose their balance: gravity. But how precisely do objects behave under the influence of gravity? We have two theoretical frameworks, both of which are still in use, depending on what we wish to calculate. Here we see an idea central to the success of science; there are no absolute truths! Usefulness is the figure of merit; if a theory can be used to make predictions that agree with experiment in certain circumstances, then as long as we understand the restrictions, we can continue to use the theory. The first theory of gravity was written down by Isaac Newton in 1687 in his *Philosophiae Naturalis Principia Mathematica* – the mathematical principles of natural philosophy, inspired at least in part by the work of our curious companion, Johannes Kepler.

A more precise description of gravity was published in 1915 by Albert Einstein. Newton's theory doesn't have anything to say about the mechanism by which gravity acts between objects, but it does allow us to calculate the gravitational force between any objects, anywhere in the Universe. Einstein's more accurate Theory of General Relativity provides an explanation for the force of gravity. Space and time are distorted by the presence of matter and energy, and objects travel in straight lines through this curved and distorted spacetime. Because of the distortion, it appears to us as if the objects are being acted upon by a force, which we call gravity. But in Einstein's picture there isn't a force; there is curved spacetime and the rule that everything travels in a straight line through it. We will encounter spacetime in much more detail in Chapter Two.

To answer the question about spherical planets, we don't need Einstein's elegant but significantly more mathematically challenging Theory of General Relativity. It is a sledgehammer to crack a nut. We'll therefore confine ourselves to Newton's simpler theory; General Relativity would give the same answer. Here is Newton's Law of Universal Gravitation:

$$F = G\,m\,M / r^2$$

In words, this equation says that there is a force between all objects, F, which is equal to the product of their masses, m and M, and inversely proportional to the square of their distance apart, r. If you double the distance between two objects, the gravitational force between them falls by a factor of 4. G is known as Newton's Constant, and it tells us the strength of the gravitational force. If we measure mass in kilograms, distance in metres and wish to know the gravitational force in Newtons, then $G = 6.6738 \times 10^{-11}\ m^3\ kg^{-1} s^{-2}$.

Newton's Gravitational Constant is one of the fundamental physical constants. It describes a property of our Universe that can be measured, but not derived from some deeper principle, as far as we know. One of the great unsolved questions in physics is why Newton's gravitational constant is so small, which is equivalent to asking why the gravitational force between objects is so weak. Comparing the strengths of forces is not entirely straightforward, because they change in strength depending on the energy scale at which you probe them; very close to the Big Bang, at what is known as the Planck temperature – 1.417×10^{32} degrees Celsius – we have good reason to think that all four forces had the same strength. To describe physics at such temperatures we require a quantum theory of gravity, which we don't currently possess in detail. But at the energies we encounter in everyday life, gravity is around forty orders of magnitude weaker than the electromagnetic force; that's 1 followed by 40 zeroes. This smallness seems absurd, and demands an explanation. Physicists speculate about extra spatial dimensions in the Universe and other exotic ideas, but as yet we have no experimental evidence to point the way. One possibility is that the constants of Nature were randomly selected at the Big Bang, in which case they are simply a set of incalculable fundamental numbers that define what sort of Universe we happen to live in. Or maybe we will one day possess a theory that is able to explain why the fundamental numbers take on the values they do.

Newton discovered his law of gravity by looking for a simple equation that could describe the apparent complexity of the motions of the planets around the Sun. Kepler's three empirical laws of planetary motion can be derived from Newton's Law of Gravitation and his laws of motion. This is why we might describe Newton's theory as elegant, in line with our discussion of quantum theory earlier in the chapter. Newton discovered a simple equation that is able to describe a wide range of phenomena: the flight of artillery shells on Earth, the orbits of planets around the Sun, the orbits of the moons of Jupiter and Saturn, the motion of stars within galaxies. His was the first truly universal law of Nature to be discovered.

The answer to our question 'why is the Earth spherical?' must be contained within Newton's equation, because the Earth formed by the action of gravity. The gravitational force is the sculptor of planets. Our Solar System formed from a cloud of gas and dust, collapsing due to the attractive force of gravity around 4.6 billion years ago. The Sun formed first, followed by the planets. Let's fast-forward a few million years to a time when the infant Sun is shining in the centre of a planet-less Solar System. Circling the young Sun are the remains of the cloud of dust and gas out of which the Sun formed, containing all the ingredients to make a planet. This is known as a protoplanetary disc. The fine details of the formation of planets are still a matter of active research, and the mechanisms may be different for rocky planets such as the Earth and gas giants such as Jupiter. For Earth-like planets, random collisions between dust particles can result in the formation of objects of around 1 kilometre in diameter known as planetesimals. These grow larger as they attract smaller lumps of rock and dust by their gravitational pull, increasing their mass, which increases their gravitational pull, attracting more objects, and so on. This is known as runaway accretion, and computer simulations using Newton's laws suggest that through a series of collisions between these ever-growing planetesimals, a small number of rocky planets emerge from the protoplanetary disc orbiting the young star.

Models of planetary formation can be checked using the telescopic observation of young star systems. In 2014 the ALMA (Atacama Large Millimeter/submillimeter Array) observatory in Chile captured a beautiful image of a planetary system forming inside a protoplanetary disc around HL Tauri, a system less than 100,000 years old and only 450 light years from Earth. A series of bright concentric rings is clearly visible, separated by darker areas. It is thought that these dark gaps are being cleared by embryonic planets orbiting around the star and sweeping up material – they are the shadow of the planetary orbits. It is interesting to note that planetary formation appears to be well advanced in this very young system. This image is perhaps a glimpse of what our Solar System looked like 4.5 billion years ago.

Rocky planets begin life as small, irregular planetesimals and evolve over time into spheres. To make progress in understanding why, we might make an observation; all objects in the Solar System are not spheres. The Martian moon Phobos has a radius of approximately 11 kilometres. It is a misshapen lump. Smaller still are the asteroids, comets and grains of dust that formed at the same time as the planets. The Comet 67P/Churyumov–Gerasimenko is less than 5 kilometres across and is an intriguing dumbbell shape. Analysis of data from the Rosetta spacecraft, in orbit around the comet at the time of writing, has shown that 67P was formed by a low-velocity collision of two larger objects. Perhaps this is a snapshot of the processes that

'FORÇA, EQUILIBRI, VALOR I SENY'

(STRENGTH, BALANCE, COURAGE AND COMMON SENSE)

previously resulted in the formation of much larger objects such as planets and moons. Smaller lumps of rock merge together under the influence of gravity, and if there is enough material in the vicinity, as there would have been early in the life of the Solar System, the objects will undergo many such collisions and grow. Why isn't comet 67P spherical?

Let's return to the human towers. What sets the maximum height of a tower? Consider an artificial situation in which the tower is a vertical stack of humans, one on top of the other. If there are only two people in the stack, then the force on the person at the base is the weight of the person above. Let's understand that sentence. What is weight? Your weight at the Earth's surface is given by Newton's equation; it is defined to be the force exerted on you by the Earth. What numbers should we put into the equation to calculate it? Your mass: *75kg*. The mass of the Earth: 5.972×10^{24} *kg*. Newton's gravitational constant, *G:* $6.6738 \times 10^{-11}\ m^3\ kg^{-1}s^{-2}$. What should we use for r? This is the distance from the centre of the Earth to the centre of you. That sounds a bit vague. More precisely, r is the distance between the centre of mass of the Earth and your centre of mass, but it's a very good approximation to simply insert the radius of the Earth into Newton's equation. This is because you are only around a couple of metres tall, and the average radius of the Earth is 6,371,000 metres, so moving your centre of mass around by a few tens of centimetres isn't going to change the calculation much.

Plugging in the numbers, Newton's equation tells us that the force on you at the Earth's surface – your weight – is approximately 736 Newtons (a force of 1 Newton produces an acceleration of 1 m/s^2 on a 1kg mass).

We now need to introduce another of Newton's laws – his third law of motion, also published in the *Principia*: To every action, there is an equal and opposite reaction. This says that the Earth exerts a force on you and you exert an equal and opposite force on the Earth. We can now understand what happens when the human towers get higher and higher. If one person stands on another's shoulders, there is a downward force on the lower person of around 730 Newtons. If another person of the same mass climbs up, the force on the person at the base doubles to 1460 Newtons. If another two people climb up to form a tower five people high, the force on the base person is 2920 Newtons, and so on. Clearly, at some point, the person at the base isn't strong enough to hold the tower up, and the whole thing will collapse. This is where the skill of the *castellers* comes in. By having a base, made up of many individuals, the forces can be distributed across the human structure, and this allows the towers to get higher before catastrophe strikes. There is clearly a trade-off; a larger base can support a larger layer above, which in turn can support a larger layer above, and so on. But a larger layer weighs more, and exerts a larger force on the layer below. The ingenious geometrical solutions to this gravitational conundrum emerge through a combination of trial and error, instinct and skill, and this is what makes the Tarragona *Castells* competition so compelling. For our purposes, it is the principle that matters. As the tower gets higher, the forces on the base increase, and ultimately a limit will be reached.

Perhaps you can see where this is leading. High human towers are more difficult to sustain because the force on the base becomes increasingly large as the mass of the tower increases. This suggests that the size of structures that rise above the surface of a planet is limited by the structural strength of the rock out of which the planet is made, and the mass of the planet, which sets the gravitational pull and therefore the weight of the structure. On Earth, the tallest mountain as measured from its base on the sea floor is Mauna Kea, on the island of Hawaii. This dormant volcano is 10 kilometres high, over a kilometre higher than Mount Everest. Mauna Kea is sinking because its weight is so great that the rock beneath cannot support it. Mars, by contrast, is a less massive planet. At a mere 6.39×10^{23} kg, it is around 10 per cent of Earth's mass and has a radius about half that of Earth. A quick calculation using the equation on page 46 will tell you that an object on the surface of Mars weighs around 40 per cent of its weight at the Earth's surface. Since Mars has a similar composition to Earth, its surface rock has a similar strength, and this implies that more massive mountains can exist on Mars because they weigh less – and this is indeed the case. The Martian mountain Olympus Mons is the highest mountain in the Solar System; at over 24 kilometres in altitude, it is close to the height of three Everests stacked on top of each other. Such a monstrous structure is impossible on Earth because of the immense weight – a result of the Earth's greater mass and therefore stronger gravitational pull at the surface.

We see that there must be a limit to the height to which a structure can rise above the surface of a planet. The more massive the planet, the stronger the gravitational pull at its surface, and the lower the height of structures that the surface can support. As the planets get more and more massive, their surfaces will get smoother and smoother because of the stronger gravity. Less-massive planets can be more uneven. We are approaching an answer to our question; we have a mechanism for smoothing out the surface of a planet, but why should this mean that planets get smoothed into a sphere?

Imagine a mountain on the surface of a planet. Let's say it is at the North Pole. Now, in your mind's eye, imagine rotating the planet through, say, 90 degrees, so the mountain sits on the Equator. Has anything changed? All the arguments about the maximum height of the mountain still apply, because the gravitational force at the surface depends *only* upon the radius and mass of the planet and the mass of the mountain. There is no reference to any angle in Newton's equation (page 46).

In more sophisticated language, we can say that Newton's law of gravitation possesses a rotational symmetry. By that, we mean that it gives the same results for the gravitational force between any two objects regardless of their orientation. This is an example of what physicists and mathematicians mean when they speak of the symmetries of an equation or law of Nature, and it means that the calculation for the maximum height of a mountain at any place on the Earth's surface must give the same answer irrespective of the position of the mountain *because* Newton's law of gravitation is symmetric under rotations. The symmetry of the law of gravity is reflected in the symmetry of the objects it forms. Gravity smooths mountains democratically, symmetrically, with the result that lumps of matter with a gravitational pull strong enough to overcome the rigidity of the substance out of which they are made end up being spherical. This is the reason why the Earth is spherically symmetric.

There is a deep idea lurking here that lies at the heart of modern theoretical physics. Thinking of things in terms of symmetry is extremely powerful, and perhaps fundamental. Consider the possibility that the laws of Nature possess certain symmetries, which are the fundamental properties of the Universe. This would be reflected in the physical objects they create. For example, imagine a Universe in which only laws of Nature that are symmetric under rotations through 90 degrees are allowed. In such a Universe, objects that remain the same under rotations through 90 degrees are created; cubes exist but spheres are forbidden. This isn't quite as crazy as it sounds. As far as we can tell, our Universe does possess a set of extremely restrictive symmetries, and the subatomic particles that exist and the forces that act between them are determined by these underlying symmetries.[6] In fact, *all* of the laws of Nature we regard as fundamental today can be understood by thinking in this way. There is certainly a strong case to be made that Nature's symmetries can be regarded as truly fundamental. The Nobel Prize-winning physicist Steven Weinberg wrote, 'I would like to suggest something here that I am not really certain about but which is at least a possibility: that specifying the symmetry group of Nature may be all we need to say about the physical world, beyond the principles of quantum mechanics.' Nobel laureate Philip Anderson wrote, 'It is only slightly overstating the case to say that physics is the study of symmetry.' Nobel laureate David Gross wrote, 'Indeed, it is hard to imagine that much progress could have been made in deducing the laws of Nature without the existence of certain symmetries … Today we realise that symmetry principles are even more powerful – they dictate the form of the laws of Nature.' The complexity we perceive when casually glancing at the Universe masks the underlying symmetries, and it is one of the goals of modern theoretical physics to strip away the complexity and reveal the underlying simplicity and symmetry of the laws of Nature.

[6] These are not symmetries of three-dimensional space, like the rotational symmetry of a cube. They are more abstract symmetries.

Returning to the task in hand, this reasoning leads to a prediction about the size and shape of planets and moons that can be checked: they should be spherical if they are large enough, and therefore massive enough, for their gravitational pull to overcome the structural strength of the rock out of which they are made. The strength of rock is ultimately related to the strength of the force of Nature that holds the constituents of rock together – molecules of silicon dioxide, iron and so on. This is the electromagnetic force; what other force could it be? There are only four forces, and the two nuclear forces are confined within the atomic nuclei themselves. Big things like planets are shaped by the interplay between gravity, trying to squash them into spheres; and electromagnetism, trying to resist the squashing. We can perform a calculation to estimate the minimum size that a lump of matter must be to form into a near-spherical shape by equating the weight of a mass of rock near its surface to the structural strength of the rocks below.[7] Our answer is approximately 600 kilometres.

We can check this by direct observation of the Solar System. Phobos fits with our prediction; with a mean radius of just over 11 kilometres and a mass of only 10^{16} kilograms, the gravitational force at its surface is far too weak to overcome the rigidity of rock and act to flatten the surface and sculpt Phobos into a sphere. At around 550 kilometres across, the asteroid Pallas is the largest known non-spherical object. Saturn's moon, Mimas, with a radius of just under 200 kilometres, is the smallest known body in the Solar System that is spherical. It is made mostly of ice, which is much easier to deform than rock – this is why it is so small and still round. Our estimate is certainly in the right ballpark.

As an important aside, 'back-of-the-envelope' estimates such as these are very important in physics; they tell us that we are on the right track, without overcomplicating things unnecessarily. We could have refined our calculation by taking into account the different compositions of different objects, and by computing the gravitational pull at different depths more carefully. We could even have tried to use General Relativity instead of Newton's laws, but we wouldn't have learnt a lot by doing so. Learning what to ignore and what to include is part and parcel of becoming a professional scientist – one might call it physical intuition. There is no precise size above which a body will be spherical; the limit depends on the object's composition; a mixture of rock and ice is easier to deform than solid rock. As a general rule, any icy moon over 400 kilometres in diameter will be a sphere. Objects made of rock need to be larger, because the gravitational force needed to deform rock is greater. If a rocky moon has an internal heat source, perhaps as a result of the presence of large amounts of radioactive material in its core or tidal heating, the body is easier to deform and may be spherical at a smaller size than would be expected for a less active object. The Solar System is full of examples of this interplay of rigidity and gravity in action, but, very roughly, we have deduced that anything with a radius in excess of a few hundred kilometres must be spherical because the gravitational forces will overcome the strength of the rock.

When gravity wins, the shape of the objects it creates reflects the underlying symmetry of the physical law, and this is why large single objects such as planets and stars are always spherical.

At larger scales, however, things change. Our nearest galactic neighbour, Andromeda, contains around 400 billion stars, formed and bound together by gravity. It is disc-shaped, not spherical. The Solar System itself is also a disc and not a sphere. Why?

Why are there discs as well as spheres in the Universe?

We argued above that planets and large moons are spheres because, if the gravitational forces are large enough to overcome the electromagnetic forces that keep matter rigid, the underlying symmetry of the gravitational force is made manifest in the objects it creates. Because there is no special direction in Newton's Law of Gravitation, there will be no special direction in the objects that it creates. This is not entirely true, however, even for planets, because they spin.

Our planet turns on its axis once every 24 hours. The spin axis marks out a special direction, which means that all points on the Earth are not the same. Someone standing on Earth's Equator is rotating at a speed of 1670 km/hour, whilst someone in Minnesota is rotating at a speed of 1180 km/hour. The spherical symmetry is broken – the two points are different. As we'll see in Chapter Two, this difference leads to observable effects such as the rotation of storm systems and the deflection of artillery shells in flight. It also leads to a very slight flattening of the Earth – the equatorial circumference is 40,075 kilometres and the polar circumference is 40,008 kilometres. The Earth is not spherical, but an oblate spheroid, because of its spin. If it were spinning faster, the Earth would be more oblate. When our Solar System formed, the spin – or more correctly angular momentum – was 'exported' outwards from the newly forming Sun, primarily through collisions and magnetic interactions in the protoplanetary disc, resulting in the system becoming flattened into a disc.

The transfer of spin outwards from the centre also results in the flattening of some galaxies; for example, Andromeda. Globular star clusters, such as the spectacular Messier 80, remained spherical because they were too diffuse for angular momentum to be transferred outwards. The shapes of objects that are bound together by gravity are therefore dependent on the amount and location of the 'spin'. For the experts, the ratio of angular momentum L to gravitational potential energy E is the figure of merit. Large L/E = disc. Small L/E = spherical.

Below: Some of the many properties that have resulted in partial loss of the symmetry in the Solar System disc.

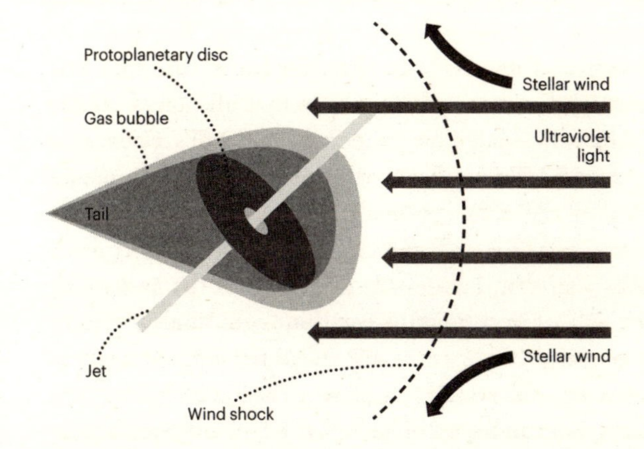

There is a very important idea hiding here. We used the term 'symmetry breaking' to describe how the presence of a spin axis marks out a particular direction, resulting in an object deviating from the 'perfect' spherical shape that reflects the symmetry of the underlying law of Nature – in this case gravity. The disc of our Solar System is less symmetric than a sphere because it only remains the same when rotated about a particular axis in space – the spin axis. The symmetry has been partially lost. We might say that the symmetry of the law of gravity that created our Solar System has been hidden by the presence of a special direction in space – the spin axis. The spin itself came from the precise details of the collapse of the initial dust cloud almost 5 billion years ago, and the distribution of the spin between the Sun and planets depended on the precise speed of collapse, the density of the protoplanetary disc and myriad other subtle details over the history of the Solar System's formation. This highlights one of the central challenges in modern science: which properties of the structures we see in Nature are reflections of the underlying laws of Nature, and which properties are determined by the history of formation or other influences? This is particularly difficult to answer when the physical systems in question are complicated. The shapes of planets, solar systems and galaxies, whilst astronomical in size, are easier to explain than the shapes of more mundane objects that we encounter every day. Let's jump from simple planets to the most complex of all physical structures – living things. By exploring the symmetries and structures of living organisms, we can further explore the idea that the shape and form of physical objects are the result of a complex interplay between deep physical principles and the history of their formation.

[7] **The smallest size of a round lump of rock and the height of the tallest mountains on Earth** Imagine a big cube of rock sitting on the surface of a second much bigger ball of rock, like a planet maybe (we are thinking of a cube for the sake of being specific but any shape will do). If the cube is too big then its weight will cause the rock underneath to fail and the cube will sink. Obviously it takes a lot of weight before rock starts to deform and give way. For granite, the maximum pressure before failure is around 130 million Newtons /m^2 (written 130 MPa), which is a little more than 1000 times atmospheric pressure. We will assume that our big ball of rock has compressive strength of around 100 MPa, and we label it using the symbol P. Now we need to know how heavy the cube is, given that its height is h. Its weight is equal to its mass multiplied by GM/R^2 (from Newton's law), where M is the mass of the big ball and R is its radius. If the density of the cube is d = 3000kg/m^3 (typical of rock) then its mass is d x h^3. The mass of the ball will likewise be M = 4/3 x 3.14 x R^3 x d (3.14 is the mathematical number pi, and we have used the formula for the volume of a sphere). For our purposes 3.14/3 is close enough to 1 as to make no difference (the goal here is to make a rough estimate, not a highly accurate computation). Together, these results mean that the weight of the cube is d x h^3 x G x 4R x d. Now, this weight bears down on the ground below, which will give way if the weight is bigger than the compressive strength of the rock supporting it, which is P x h^2. In other words, the ground will give way under the cube if h^3 x G x 4R x d^2 is bigger than P x h^2. This implies that h must be smaller than $P/G/4/R/d^2$. If this maximum value for h is less than 10 % of the radius of the ball, the surface of the ball will not be too much deformed from spherical by the cube. (i.e. the cube will be a small bump on the surface of a bigger ball). Putting h/R = 0.1 tells us that the planet's radius R must be bigger than the square root of $P/G/4/d^2$/10 %. Putting in the numbers gives a radius equal to just over 600 kilometres. This number should not be taken too literally, because we used typical numbers for the density and the compressive strength and these will vary across the variety of planets, asteroids and comets. But that should not detract from what we have achieved. Our calculation is telling us that lumps of rock larger than about 600 kilometres in radius will tend to look pretty smooth because big structures on their surface will tend to sink down and be absorbed. While we are at it, we can quickly go ahead and estimate the size of the biggest mountains on Earth and Mars. We have already worked this out above. The maximum size of a cubic mountain on Earth would be $P/G/4/R/d^2$. On Earth, the combination GM/R^2 (where M is the mass of the Earth and R is its radius) is called the acceleration due to gravity, g, and it is close to 10 m/s^2. This means that our cubic mountain would sink if it were taller than P/d/g, which is around 3.3 kilometres. If the mountain is cone shaped instead of cubic then this number increases by a factor of 3 to around 10 kilometres, which is very close to the height of the largest mountains on Earth. On Mars, the surface gravity is around 40 % that of the Earth, which means that its tallest mountains should be more like 10 kilometres/40 % = 25 kilometres high, which is the height of Olympus Mons.

Why does life come in so many shapes and sizes?

The competition between the force of gravity and the electromagnetic force is responsible for smoothing the surface of planets and moons into spheres and limiting the maximum size of mountains on their surfaces. One of the central ideas in this book, which we will expand on in Chapter Three, is that there is no fundamental difference between inanimate things, such as planets, and living things, such as bacteria or human beings; all objects in the Universe are made of the same ingredients and are shaped by the same forces of Nature. We should therefore expect to see limits on the form and function of living things imposed by the laws of Nature. Basic physics is not the only driver of the structure of organisms, of course; there is also the undirected hand of evolution by natural selection, which moulds living things over time in response to their changing environment, their interaction with other living things, and the myriad available environmental niches. This creative interplay between the relentless determinism of physical laws and the seething, infinitely intertwined, ever-shifting genetic database of life on Earth is beautifully captured in Darwin's closing lines of *On the Origin of Species*;

> *'There is grandeur in this view of life, with its several powers, having been originally breathed into a few forms or into one; and that, whilst this planet has gone cycling on according to the fixed law of gravity, from so simple a beginning endless forms most beautiful and most wonderful have been, and are being, evolved.'*

Another of our recurring themes is a celebration of the energetic curiosity of the early scientists. There is a breathless lyricism in their descriptions of ideas that remains relevant and essential; yet their presentation seems somehow unencumbered by the more serious and confining demands of modern professional science. There are great writers from the modern era who capture the logic, clarity and wonder of science – Richard Feynman, Richard Dawkins and Carl Sagan spring immediately to mind – but there is something exhilarating in seeing science evolving in words. The limits of the Renaissance authors are so often coterminous with the limits of all human knowledge that the investigations on the page are near real-time explorations rather than reminiscences from a well-trodden intellectual road. Perhaps this is what gives the old masters' writings such exhilarating intellectual pace.

Just as the force of gravity limits the maximum size of Earth's mountains, so it limits the range of forms that natural selection can create, restricting the overall size of organisms that live on its surface. Four hundred years ago, Galileo Galilei explored the factors that define how big an animal can be; in common with Kepler and his snowflakes, he was operating at the edge of knowledge and ahead of his time. *Discourses and Mathematical Demonstrations Relating to Two Sciences* was Galileo's final book, written whilst under house arrest and published in 1638 by the Dutch publisher Lodewijk Elzevir, because no country in the grip of the Inquisition would touch it. Any scientist reading this book will recognise the name: the Elsevier company, which took the publisher's name, is today a leading scientific publisher. Galileo's book is written in the style of a conversation between three men, Simplicio, Sagredo and Salviati, who each represent the author at a different age, and with a different level of knowledge. The characters wander from question to question during a conversation lasting four days, discussing and debating each subject before moving on to the next. The book has something of the voyeuristic pleasure of overhearing a conversation on a park bench – albeit in a park frequented by unusually thoughtful individuals. Galileo covers large swathes of the physics of the day, including a critical look at Aristotelian physics, accelerated motion, the motion of projectiles and the nature of infinity. His investigations also turned to the strength of materials and the limits placed on the size and form of structures, both animate and inanimate, by the laws of Nature.

From what has already been demonstrated, you can plainly see the impossibility of increasing the size of structures to vast dimensions, either in art or in Nature. Likewise the impossibility of building ships, palaces or temples of enormous size in such a way that their oars, yards, beams, iron-bolts and, in short, all their other parts will hold together. Nor can Nature produce trees of extraordinary size, because the branches would break under their own weight; so also it would be impossible to build up the bony structures of men, horses or other animals so as to hold together and perform their normal functions if these animals were to be increased enormously in height. For this increase in height can be accomplished only by employing a material which is harder and stronger than usual, or by enlarging the size of the bones, thus changing their shape until their form and appearance suggest a monstrosity.

Galileo states, for the first time, the relationship between volume and area, known today as the square–cube law; as an object grows in size, the volume grows faster than the surface area. Consider the example of a cube of sides measuring 2cm. The surface area is 6 x 2 x 2 = 24cm^2. The volume is 2 x 2 x 2 = 8cm^3. If we double the length of the sides, the surface area is 96cm^2 and the volume is 64cm^3. Double the length of the sides again and the surface area increases to 384cm^2 whilst the volume is 512cm^3. And so on.

This means that, as animals get larger, their volume, and therefore their mass, increases more rapidly than their surface area and the cross-sectional area of their bones. The consequence of this is that animals can't simply be 'scaled up' in size. A mouse can't be expanded to the size of an elephant because its skeleton would give way; that's why an elephant has thicker legs relative to the rest of its body than a mouse. This ultimately places a fundamental limit on the maximum size of living things on land; the structural strength of bone, or wood in the case of trees, limits the mass of the organism in the same way that the structural strength of the rocks of the Earth's crust limits the size of a mountain. On Mars, elephants could have thinner legs.

Galileo realised there was an exception to this rule. Whereas gravity imposes a limit to the size and shape of animals on land, the constraints placed on living things by physical laws are different in water. Marine animals float, which means the effects of gravity are not relevant. With the necessity for strong bones to support their weight removed, their forms are freed from this particular constraint.

Here is how Simplicio, Sagredo and Salviati put it, from their metaphorical park bench. I can't help but hear them as a sort of three-way Renaissance version of Pete and Dud…

Simplicio: *This may be so; but I am led to doubt it on account of the enormous size reached by certain fish, such as the whale which, I understand, is ten times as large as an elephant; yet they all support themselves.*

You read that in a Dagenham accent, didn't you?

Simplicio: *A very shrewd objection! And now, in reply, tell me whether you have ever seen fish stand motionless at will under water, neither descending to the bottom nor rising to the top, without the exertion of force by swimming?*

Simplicio: *In aquatic animals therefore circumstances are just reversed from what they are with land animals inasmuch as, in the latter, the bones sustain not only their own weight but also that of the flesh, while in the former it is the flesh which supports not only its own weight but also that of the bones. We must therefore cease to wonder why these enormously large animals inhabit the water rather than the land, that is to say, the air.*

Sagredo: *I am convinced and I only wish to add that what we call land animals ought really to be called air animals, seeing that they live in the air, are surrounded by air, and breathe air.*

Salviati: *I have enjoyed Simplicio's discussion, including both the question raised and its answer. Moreover I can easily understand that one of these giant fish, if pulled ashore, would not perhaps sustain itself for any great length of time, but would be crushed under its own mass as soon as the connections between the bones gave way.*

Freed from the tyranny of gravity, aquatic animals can be larger than their land-based cousins, but they don't have complete freedom from the laws of physics.

Every winter the warm waters of Florida are home to one of Nature's apparently less elegant shapes. The caveat is important, because the clumsy-looking manatee is as well adapted to its environment as the most aesthetically refined butterfly. The West Indian manatee is the largest living example in the Sirenia order of wholly aquatic, herbivorous mammals. A less-than-taxonomically accurate but nonetheless accurate image can be conjured by imagining a 4m-long aquatic cow with no legs, unhurriedly grazing on the sea grasses that grow in the slow-moving waterways along the Floridian coast.

During the summer months the manatee roam as far north as Massachusetts, but as the seasonal temperatures fall they must return to warmer seas. They are unable to survive in waters below 20 degrees Celsius for long. The need for warm winter waters drives the manatees to congregate in large groups around the warm springs that dot the Florida coast, where temperatures remain above 22 degrees Celsius all year round. They also take advantage of human activity, gathering in the outflows of power plants near Apollo Beach and Fort Myers. The manatee is a strange animal indeed; it is more closely related to an elephant than to the other marine mammals – they share a common ancestor around 60 million years ago, not long after the dinosaurs became extinct. The ancestor may have looked like the modern-day hyrax, which at around 50cm in length looks nothing like an elephant or a manatee; 60 million years is plenty of time for the un-directed tinkering sieve of natural selection to sculpt an animal to take advantage of an environmental niche.

The elephant's niche is to be the biggest land animal, which undoubtedly gives it an advantage against predators, but it also displays the anatomical evidence of a tussle with gravity. As dictated by the square-cube law, the elephant has evolved with exceptionally thick legs to support its substantial weight. There is also the matter of cooling; heat escapes from an organism through its surface. As the volume of the animal increases, so does the amount of heat it generates, but its surface area decreases in proportion, according to the square-cube law. This presents a problem for a land-dwelling animal, and the elephant has solved it by developing an ingenious cooling system – its big ears.

The manatee filled a different niche. The transition from a coastal land-dweller to an aquatic mammal saw their front limbs evolve into flippers, although they still possess their ancient finger-bone structure and fingernails. The rear limbs have become a giant paddle-shaped caudal fin, a gradual evolutionary change wonderfully documented in the fossil record. The limbs of the ancient ancestor that grew thick to resist gravity in the elephant have become streamlined to allow the manatee to swim at up to 12 km/hour. The manatee can dive deep, for up to twenty minutes at a time, but being an air-breathing mammal it must surface for air eventually. Its time under water is maximised by slowing down its heartbeat and metabolic rate, reducing the need for oxygen; but this is where biology comes into conflict with physics. A low metabolic rate means limited heat production, and water is an extremely good conductor of heat away from the body, so there is a danger of becoming too cold. The compromise solutions discovered, naturally, by natural selection, are to get bigger, which reduces the surface-area-to-volume ratio and therefore decreases the rate of heat loss per unit volume, and to get spherical (see illustration, below).

Below: This graph shows how the surface area decreases for rounder shapes and the surface-area-to-volume ratio decreases as the volume increases.

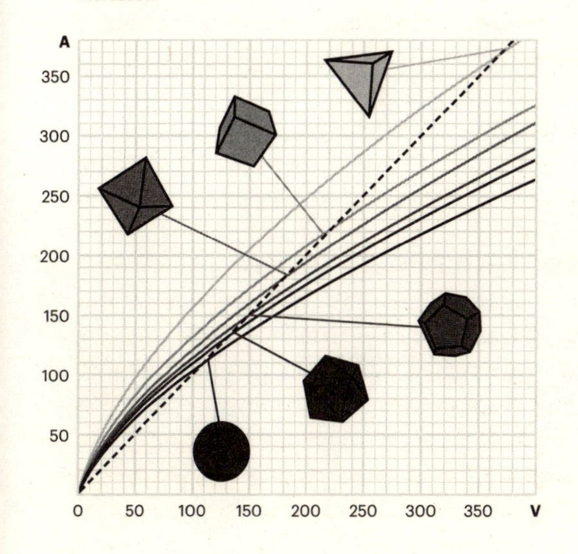

This is a beautiful example of a naturally occurring shape reflecting a deeper mathematical reality. The sphere is the three-dimensional shape with the lowest surface-area-to-volume ratio. If you want to generate lots of heat by having a large volume, but lose as little through your surface as possible, you'll be spherical – and the manatee is the most spherical mammal on Earth. What a wonderful thing to be – unless you are an astronomer. The astronomer Fritz Zwicky is credited with calling a group of his colleagues Spherical Bastards, because they are bastards, whichever way you look at them. Which brings us nicely back to the subject of symmetry. If a physicist designed a manatee it would be spherically symmetric. Symmetrical shapes such as planets tend to be the result of the action of symmetrical laws of Nature, unless there are reasons for the symmetry to be broken. There are no perfectly symmetric large organisms in biology. Why?

Symmetry and symmetry breaking in biology

Leonardo da Vinci's 'Vitruvian Man' is perhaps the most famous drawing of the human form in history. It depicts a man in two superimposed positions within a circle and a square. The proportions are carefully calculated in an attempt to represent the underlying perfection of Man and to link him directly to the Universe. Da Vinci was inspired by one of the great classical works, *De architectura*, written by the Roman architect Vitruvius. The relationship of the human form to a circle and square reflects ancient ideas – dating back to Plato, Pythagoras and earlier mystic traditions – which attempted to forge a link between Nature and geometry. Kepler's early work on the motion of the planets was firmly rooted in this tradition, and he only jettisoned the idea that the motion of the planets could be described in terms of the perfect 'Platonic' solids when the data forced him to conclude that planets actually move in elliptical orbits rather than circular ones. It is interesting to reflect on the fact that the explanation for the motion of the planets is more elegant and beautiful than Kepler's hoped-for geometrical perfection. As we've seen, the motion of all the planets and moons in the Solar System, and indeed every solar system in the Universe, can be described by the application of Newton's laws of motion and Universal Gravitation; a profound simplification that would surely have appealed to Kepler, and to Plato before him, because Newton's laws do embody a 'perfect' spherical symmetry, which is hidden but still evident in the structures it creates.

Similarly, from a modern perspective, the interesting thing about the human form is not its symmetry but its partial symmetry. As we've discussed, a sphere is a perfectly symmetrical three-dimensional form, in that it looks precisely the same from any angle. Humans, in common with most, but not all, large animals, have a head at one end and an anus at the other, and they have very different functions in the majority of sensible individuals. Humans exhibit what is known as bilateral symmetry, at least externally. This means that there is a symmetry axis, running downwards from our heads to our anus, about which we exhibit mirror symmetry; we have left and right halves. Why?

This book is based on a television series. Part of the challenge of making a documentary that aims to explain or describe abstract scientific concepts is to work out what to point the camera at. The days when an audience would watch a middle-aged academic in a tank top with unkempt hair and a pencil are long gone. I regret this, because I was born in 1968, I have naturally bizarre hair and a big pencil case. Someone at the BBC had a very good idea for illustrating the range of symmetries of living things; on the island of Marado off the southernmost coast of South Korea they found a community of free-diving grannies.

For centuries the women in this region of Korea have been free divers, harvesting the seabed for valuable abalone, conch, sea urchin and octopus. Their bounty is so important that Marado has become one of the world's rare matriarchal societies; the women traditionally work in the ocean, while the men bring up the children and run the home. In the cold waters of the East China Sea, becoming a Haenyeo or 'sea woman' is not easy. Girls start their training as young as 11, but once proficient many continue to dive for their whole working lives. It is not clear how the tradition began, but from a biological perspective it makes sense; women are better adapted for work in cold water than men. Before the arrival of wetsuits the Haenyeo would have dived with little protection, and the additional body fat in the female body would have been an advantage. The divers often remain in the water for several hours, free diving for up to two minutes each time to depths beyond 20 metres. Today, the average age of a Haenyeo is 65, and some of the divers are in their 80s. This demographic cliff reflects a changing world. There are now fewer than 2000 Haenyeo; once there were 50,000. The riches of the deep appeal less to the twenty-first-century generation, although the Haenyeo are still highly respected in Korean culture and tourists come from far and wide to watch the women dive and to eat the seafood they catch.

The oceans are home to the most visible assortment of different body plans. Starfish, sea anemones and jellyfish exhibit radial symmetry, at least superficially; they have a top and bottom, which can be defined by the position of their mouth, but they do not have 'left' and 'right' sides. A jellyfish has continuous rotational symmetry about a central axis, whereas a five-armed starfish looks the same if it is rotated through an angle of 72 degrees. The octopus exhibits bilateral symmetry as we do: mirror symmetry around a central axis, notwithstanding the fact that it has eight arms. Some sponges have no symmetry at all. What is the origin of this great range of symmetries in living things? This is a very good question, and it is also a topic of ongoing research. The oldest widely accepted fossil evidence of bilateral symmetry dates back 555 million years ago to a slug-like creature called Kimberella, a rock-dwelling organism that has been found in both Southern Australia and most numerously near the White Sea in Russia. This is just before the Cambrian explosion, the great diversification of life that follows the earliest evidence of multicellular organisms in the fossil record during the Ediacaran period 600 million years ago.

The most widely accepted view is that having bilateral symmetry confers an advantage over radial symmetry because it enables organisms to move more efficiently. Think of the shape of a shark. It is sculpted like a submarine for good reason; it faces the same engineering challenge of moving quickly and efficiently through water. We don't build high-speed radially symmetric submarines; they all look like sharks. As life began to sense the world and move around in ever more complex ways, it was a body plan with a left and right and a top and bottom that could best house a central nervous system and provide the agility needed in the emerging world of predator and prey.

There is, however, another possibility. Jellyfish also exhibit bilateral symmetry, but it is internal. From an evolutionary perspective, this is important. It may suggest that bilateral symmetry initially evolved to improve the internal functioning of organisms; it allows, for example, for an efficient separation of the gut and respiratory systems. Furthermore, the genes responsible for the development of bilaterally symmetric animals are also found in jellyfish and sea anemones. This is taken as evidence that the common ancestor may have possessed bilateral symmetry, and the external radial symmetry we see in some complex multicellular organisms was a later evolutionary development.

This is yet another example of the wonderful pace and ever-shifting nature of science. Yesterday's textbook explanation can become tomorrow's historical curiosity, and this is precisely as it must be if our knowledge of the natural world is incomplete and continually growing. Progress in science implies that we understood less yesterday than we will tomorrow. That said, I think it is unarguable, and wonderful to consider, that we can trace our ancestry back through geological time to a period in Earth's history some half a billion years ago, when we shared a common ancestor with all the multicellular animals present on Earth today, and it would have looked something like the Kimberella.

Below: The morphology diagram shows how the structure of snowflakes varies with temperature and humidity.

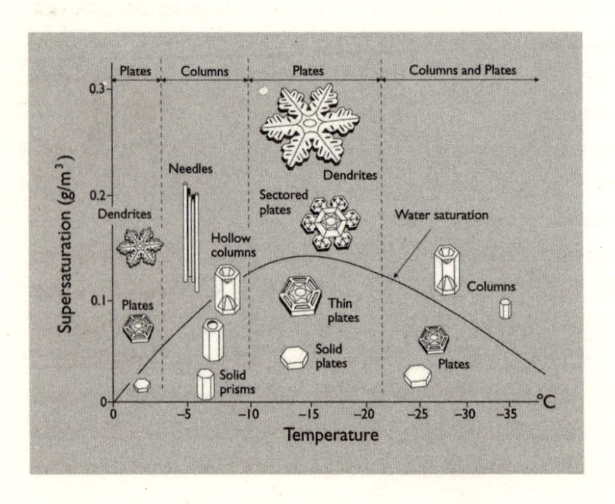

The Universe in a snowflake

Let's finish where we began, bringing together everything we've discovered to answer Kepler's question about the origin of the individuality and collective symmetrical beauty of snowflakes. We will follow the formation of a snowflake from its beginnings in a high cloud to its gentle arrival on the ground. Snowflakes are formed by water vapour condensing directly into ice. They are not frozen raindrops; they are crystals that grow steadily larger as they journey through the clouds.

The structure of ice crystals at the temperatures and pressures that we find on Earth is shown in the illustration opposite. The 104.5-degree bond-angle of a free water molecule – a consequence of the laws of quantum theory – is the reason for a hexagonal crystalline structure, which in turn is the underlying reason for the six-fold symmetry of snowflakes. The hexagons are clearly visible in Snowflake Wilson's photographs in the plate section. The snowflakes are imperfect shadows of a more 'perfect' form – the ice lattice; itself a consequence of the structure of the water molecule, which is a physical manifestation of the underlying fundamental laws of Nature that created it – the quantum theories of the strong, weak and electromagnetic forces. When you look at a snowflake, you are seeing the primal structure of our Universe.

And yet, despite the underlying simplicity, each snowflake is different. Why? Because of their individual formation histories. As we saw for planets, galaxies and grannies, simple laws of Nature can sculpt an infinity of forms because the initial conditions and histories of formation are never precisely the same. The symmetries of the laws are obscured by history, and it is the job of the scientist to see through the distorting lens of history. A clue as to how this can be done for snowflakes can be found in what is known as the morphology diagram, shown in the illustration on page 70. The morphology diagram shows how the structure of snowflakes varies with temperature and humidity.

The vertical axis of the diagram shows the humidity: the moisture content of the clouds within which the snowflakes form. The horizontal axis shows the temperature. Large, fluffy snowflakes with lots of intricate branches form at high humidity and temperatures between about -10 degrees Celsius and -20 degrees Celsius. At lower temperature and humidity snowflakes are small unbranched hexagons. Higher temperatures lead to needles and prism shapes. For complex, intricate snowflakes, humidity needs to be high. The diagram provides a clue as to how snowflakes can be similar and yet individual. By plotting the data in this way, we see that different patterns of crystal growth are favoured by different conditions and histories of formation.

To better comprehend this, we need to understand how the snowflake crystals grow. The process by which the geometry of the water molecules is transferred to the snowflake is known as faceting. A small ice crystal in a cloud grows because other water molecules bump into it and stick to it through hydrogen bonds. Faceting occurs because rough, uneven bits of the crystal have more available sites for water molecules to bind to; smooth bits, on the other hand, have fewer. This means that rough regions of the initial crystal will grow faster than smooth regions, and become smoother as the jagged spots are filled in. Faceting produces flat, hexagonal prism shapes like those labelled 'solid plates' in the morphology diagram. The plates are flat because water molecules are more likely to bind to the rectangular thin edges, which are known as prism facets, than to the hexagonal top and bottom surfaces, which are known as basal facets. In low-humidity conditions this is the dominant method of growth, which is why snowflakes can remain broadly hexagonal with few intricate branches. If you look back to Snowflake Bentley's photographs in the plate section, the snowflake in the top right-hand corner, labelled 780, is of this type.

Complexity arises from another form of crystal growth called the branching instability. If a bump forms on the crystal surface, the tip of the bump is slightly more likely to accumulate water molecules because it sticks out further into humid air. This causes the bump to grow rapidly, which is why it is referred to as an instability. Branching competes with faceting for a limited supply of water molecules, and it is this competition that leads to the complexity of snowflakes. The corners of the hexagonal prisms are subject to the branching instability – they tend to grow faster than the flat sides, causing them to become concaved. This is resisted by faceting, because there are more sites available for bonding in the centre of the concaved surface between the edges. More water molecules are available at the points, but water molecules are more likely to stick to the centre of the resulting curves. The two processes compete with each other for water molecules. If the air is humid and there is a plentiful supply of water molecules, branching dominates because the rate of growth of the instabilities outstrips the rate at which water molecules diffuse down to the faceted surface below; the corners of the hexagonal prisms therefore tend to grow rapidly, producing intricate, star-like branches. If the air is less humid, the growth rate of the branches falls below the diffusion rate and faceting dominates; the crystal evolves towards a smoother, simpler shape.

As the snowflake grows, it will pass through many different regions in a cloud and experience different conditions, passing through more humid and less humid air, and through regions of differing temperature. Each of these different regions will favour a different type of growth; sometimes faceting will dominate, and other times branching will win out. Look again at Snowflake Bentley's images. You can read the history of each snowflake from the inside out; they all begin with little hexagons; when they are small, faceting always wins. If they enter humid air, branching drives an explosion of intricacy. They may drift into a less humid region and the smooth, faceting growth reasserts itself. This is the reason why every snowflake is different. Each one follows a unique path through the clouds, and every detail of this path is written into its structure. The snowflakes retain an element of the underlying symmetry of the crystal because conditions do not vary in clouds over distances of a few centimetres, which is the size of a snowflake. Each corner therefore experiences precisely the same conditions, which lead to the same structural growth. If one side of a snowflake experiences a different history to the other – perhaps it is involved in a collision – then the symmetry of the snowflake is lost. There are, of course, many snowflakes that reach the ground in a battered, asymmetric state, but we don't take pictures of those!

As a physicist, I have to observe that snowflakes are four-dimensional objects; their structure can only be understood with reference to their history, and their history is encoded visibly into their structure. You can read a snowflake like a history book. Precisely the same observation can be made about living things. It is impossible to understand the structure of a manatee unless you understand its evolutionary history. Why does a manatee have finger bones embedded in its flippers? Because they evolved from the legs of a small land-dwelling ancestor. Living things are a snapshot, a temporal shadow of a much grander four-dimensional story; they encode the entire history of life on Earth, stretching back four billion years, into their structure. No wonder they are complex and difficult to understand. Every twist and turn of history is faithfully recorded.

The interplay between the laws of Nature, which are simple and deeply symmetric, and history, which is long and messy, produces the complex world we inhabit. The triumph of modern science is that we can separate the two, and this has led to discoveries of overwhelming importance. The seeds of this approach are clearly visible in the writings of Kepler, all those years ago. 'Since it always happens, when it begins to snow, that the first particles of snow adopt the shape of small, six-cornered stars, there must be a particular cause; for if it happened by chance, why would they always fall with six corners and not with five, or seven … ?' he asks. And there it is: Nature is beautiful, deep down, and we want to glimpse that underlying beauty. Let's not guess. Let's not make something up. Let's think, observe, experiment, pay attention, look for similarities and differences across the natural world and try to understand them. Most of all, let's be comfortable, delighted, exhilarated when faced with the unknown and devote our time to exploring the infinite territory beyond. There are treasures beyond imagination in the simplest things, if we care to look closely.

YOU CAN READ A SNOWFLAKE
LIKE A HISTORY BOOK.
PRECISELY THE SAME
OBSERVATION CAN BE MADE
ABOUT LIVING THINGS.

MOT

ION

Somewhere in spacetime

Do you remember a perfect summer's day? Not precisely where or when, but a moment of languid warmth, pepper scents, itchy grass, airborne seeds and brushing insects. Great artists are able to summon the tangled past into vivid present experience because memories are indelible. Claude Monet captured notes of a thousand childhood summers in his *Coquelicots* (Poppies). Depicting a rural landscape near the village of Argenteuil, it remains one of Monet's most loved and most recognised paintings.

Monet painted a series of similar pictures during the summer and autumn of 1873. The precise date and time of the scene are unknown, but the poppy fields around Argenteuil are in full bloom in late May and early June, and the bright sky and lack of shadow suggest that the scene was set at around noon. Let's take artistic licence of our own, though, and label the moment when a little boy ambled through the poppies with his mum, and Monet placed a carefully considered dab of red paint on his canvas. Let's say it was noon, 26 May 1873, in a field close to the village of Argenteuil, France. Almost 150 years later, that day has slipped from living memory and exists only in Monet's painting. The place is still there, but the moment is gone. Whimsical common sense, you may say, but is it correct?

In 1905, Albert Einstein published his Theory of Special Relativity, which contains the famous equation $E=mc^2$. I take great comfort in the fact that there is such a thing as a famous equation; it allows me to imagine that I glimpse a flicker of intellectual depth illuminating the all-enveloping darkness of popular culture. Special Relativity deals with moments, or more precisely events. An event is something that happens at a particular location in space and at a single instant in time. Monet's dab of paint on the canvas is an event; it has a location

‘SUDDENLY, FROM BEHIND THE RIM OF THE MOON, IN LONG, SLOW-MOTION MOMENTS OF IMMENSE MAJESTY, THERE EMERGES A SPARKLING BLUE AND WHITE JEWEL, A LIGHT, DELICATE SKY-BLUE SPHERE LACED WITH SLOWLY SWIRLING VEILS OF WHITE, RISING GRADUALLY LIKE A SMALL PEARL IN A THICK SEA OF BLACK MYSTERY. IT TAKES MORE THAN A MOMENT TO FULLY REALIZE THIS IS EARTH . . . HOME.’

– EDGAR MITCHELL, APOLLO 14, IN ORBIT AROUND THE MOON, FEBRUARY 1971 and there is a time that it happened. Einstein's theory tells us how we should measure the distance between events and how to think about their connection to each other. It is a theory of space and time.

What is time? Everyday experience tells us that time is something that passes, measured by the ticking of a clock. If everyone synchronises clocks, and those clocks are mechanically perfect, we might expect that everyone will agree on the time for evermore. We exist in the present, and we are comfortable defining the present moment as 'now'. Since we all agree on what time it is, therefore we must agree on what 'now' means. This implies that the past is gone, fading in memory as a canvas dims with age, and the future is yet to come.

What is space? Space feels like the arena within which things happen; a giant box containing Earth, Moon, Sun, planets and stars. Two people could measure the distance in space between Earth and Moon at an agreed time on perfectly synchronised clocks with perfectly calibrated rulers, and they would agree.

In this chapter, we will follow Einstein in discovering that the obvious statements in those last two paragraphs are wrong: we will discover that time and space are not what they seem. This will lead us to consider the startling possibility that Monet's magical summer's day may have an existence beyond his ageing canvas. As the great physicist and mathematician Hermann Weyl wrote, 'The objective world simply is, it does not happen. Only to the gaze of my consciousness, crawling along the lifeline of my body, does a section of this world come to life as a fleeting image in space which continuously changes in time.'

But we must start at the beginning, and think carefully about how concepts such as distances in space and intervals in time are treated in physics. The first steps towards the modern understanding of space and time were arguably the first steps along the road to modern science itself. Our story begins in the seventeenth century with Galileo, Newton, and the systematic study of the motions of the planets and Earth's place in the Solar System.

'ALAS I HAVE LITTLE MORE THAN VINTAGE WINE AND MEMORIES.'

– UNCLE MONTY, WITHNAIL AND I, CAMDEN, 1987

Life on a spinning, orbiting planet

The study of motion has a long and controversial history that stretches back many thousands of years. At first sight it is hard to imagine how the study of motion could ever be controversial; it seems like such a basic thing. The origin of the controversy was, in part, due to the fact that we live on a spinning planet that is hurtling around the Sun. That statement was still problematic in Newton's time, partly for well-known theological reasons but also because it really doesn't feel as if we are moving. Common sense informs us that we are standing still, and common sense of the 'I might not know much about science but I know what I think and feel' variety has a profoundly negative effect on public discourse in the twenty-first century – never mind in the seventeenth. If the feeling that we are standing still at the centre of the Universe on an immovable planet really were reliable, Galileo and many others would have been saved a lot of bother.

It is sometimes the case that remarkable ideas become so embedded in culture that they cease to feel remarkable simply because they are familiar. The motion of the Earth around the Sun and the sheer hidden violence of the celestial dynamics that Nature conspires to conceal from us is an excellent example of this conundrum. Most of us give very little thought to what's actually happening to the ground beneath our feet because we've been taught to hold difficult concepts in our heads with reckless intellectual abandon. An educated person probably knows that we're all walking around on the surface of a sphere of equatorial circumference of 40,000 km and mass 6 thousand million million million tonnes, spinning around an extravagantly tilted axis once every 24 hours, and that the whole vast spinning thing is barrelling around the Sun at close to 30 kilometres per second in order to make it around the 940-million-kilometre orbit every year. Such a person probably doesn't find it amazing that we don't notice this on a day-to-day basis. It's dizzying.

The reason why we don't notice is a deep one, and to appreciate it we need to explore precisely what we mean by motion. Before the seventeenth century it was widely believed that things move when they are pushed and stand still when they are left alone. Aristotle is usually credited with the expression of this intuitive view, based on the idea that everything that happens must have a cause. Since motion is something that happens, involving a change in the position of an object over some interval of time, it must have a cause. If the cause is removed, the motion should stop. Intuitively reasonable perhaps, but not correct.

It is true that if you push an object along a table, it moves, and if you stop pushing it, it stops. That is because friction between the object and the table slows it down. If there is no friction and you give the object a push, it will carry on moving until you give it another push. This is known as the principle of inertia, and it is a remarkable thing when you think about it.

The incomparable Nobel Prize-winning physicist Richard Feynman described how his father introduced him to the concept of inertia when he noticed something whilst playing with a toy wagon and a ball.

'"Say, Pop, I noticed something. When I pull the wagon, the ball rolls to the back of the wagon, and when I'm pulling it along and I suddenly stop, the ball rolls to the front of the wagon. Why is that?"

'"That, nobody knows," he said. "The general principle is that things which are moving tend to keep on moving and things which are standing still tend to keep standing still, unless you push them hard. This tendency is called inertia, but nobody knows why it's true."

'Now, that's a deep understanding. He didn't just give me the name.'

I like this because it illustrates something important. There are some questions about Nature that have the answer 'because that's the way our Universe is'. There have to be answers like this, because even if we knew how to derive all of the laws of Nature from first principles, we'd still need to know what those principles are. The law of inertia, as expressed by Feynman's dad, is one such principle as far as we know. One of the most difficult things in modern physics is to find out which properties of the Universe are truly fundamental and which follow from a deeper principle or law. This book is all about asking 'Why?' Sometimes, the answer is 'because it is'. This may be wrong – there may be a deeper reason for something that we haven't yet discovered, but it isn't a superficial answer.

Isaac Newton expressed the principle of inertia in the first of his three laws of motion, which he published in 1687 in *The Principia*

‘ABSOLUTE, TRUE AND MATHEMATICAL TIME, OF ITSELF, AND FROM ITS OWN NATURE FLOWS EQUABLY WITHOUT REGARD TO ANYTHING EXTERNAL …’

– ISAAC NEWTON

Mathematica. Virtually everyone today can recite it word for word, or at least remembers a time at school when they could:

'Every object continues in a state of rest or uniform motion in a straight line unless acted upon by a force.'

There is a subtlety here. If we are to say that an object is moving, then we have to answer the question 'relative to what is the object moving?' Newton certainly thought about this question, and almost got the answer right. His writings on the subject are illuminating, and go to the heart of questions about the nature of space and time, and how they are linked to motion. Newton stated the assumptions behind his laws clearly.

'Absolute, true and mathematical time, of itself, and from its own nature flows equably without regard to anything external ...'

That's the intuitive view that time ticks along, and everyone agrees on the rate at which it ticks.

'Absolute space, in its own nature, without regard to anything external, remains always similar and immovable ... Absolute motion is the translation of a body from one absolute place into another.'

This is Newton's assertion that there is some sort of giant box within which everything happens. We can go a little further and imagine a series of grid lines crisscrossing the box, against which we can mark the position of anything in the Universe. We could then define absolute motion as being motion relative to this universal grid, which we assert to be standing absolutely still in absolute space. This giant grid is an example of what we'll call a frame of reference. In order to define absolute motion, Newton is assuming that there is a very special frame of reference: the frame corresponding to the universal grid, at rest with respect to absolute space, against which all motion is measured.

Then, wonderfully, Newton makes a further observation;

'... but motion and rest, in the popular sense of the term, are distinguished from each other only by point of view, and bodies commonly regarded as being at rest are not always truly at rest.'

Newton is saying that it is impossible to determine whether or not an object is 'actually' in motion in a straight line, or 'actually' standing still. We might not be 'truly at rest', as he puts it, but we can't tell. This is the reason why we don't feel as if we're moving around the

Sun while we are standing on the surface of the Earth; on minute-to-minute timescales, we are almost travelling at constant speed and approximately in a straight line. Newton was correct in noticing that if this is the case we won't feel as if we are moving; indeed, we are at liberty to claim that we are at rest, even though we might not be, in his language, 'truly at rest'.

Let us make an apparently philosophical aside that has extremely important consequences for the development of Einstein's theory of relativity. If it's impossible to decide whether or not we are moving, even in principle, then what use is the concept of absolute space? Is there, in reality, no special frame of reference against which all motion can be judged? Shouldn't we just jettison the idea? Yes, that is correct, we should, but Newton never did. The wonderful thing is that his laws of motion do only deal with relative motion, and do not rely on his assumption about the existence of a special frame of reference against which all motion should be calibrated. He got the equations right, but then saddled their interpretation with the unnecessary philosophical baggage of absolute space. All of this might seem like pedantry without relevance, but it isn't. The redundant but comforting idea that space is the fixed arena within which 'stuff happens' is positively harmful to our understanding of Nature. Jettisoning it allowed Einstein to construct an entirely new theory of space and time, which delivers a more accurate description of the natural world than Newton's laws.

This does not mean that we want to jettison the concept of a frame of reference – far from it! I've realised something about physics during my years of trying to understand it for myself and explain it to others. Truly deep concepts often sound like utter pedantry. This is one of the few similarities between physics and philosophy. Our careful introduction to the idea of frames of reference is a good example; it may seem that we've been almost too careful, but we'll need to take care if we are to understand the somewhat cryptic comments we've made so far about the implications of Einstein's Theory of Special Relativity. With that in mind, let's take a brief diversion to explore frames of reference in more detail. The effort will be worth it.

An important aside: frames of reference

We can imagine erecting a set of grid lines that span the Universe, just as Newton did. The positions of objects can then be measured with reference to the grid. This grid represents a frame of reference.

Reference frames are more than an interlocking set of rulers, however. We also need to measure time. Let's also imagine an array of identical clocks scattered across the Universe. All of the clocks sit at fixed positions with respect to the grid. We can now go ahead and measure where and when an event happened; it happened at some position in space (we can use the grid to record precisely where) and at some particular time (we can use the clock adjacent to the event to record precisely when).

It isn't overstating things to say that the whole of physics can be reduced to understanding the relationships between events. This is why we are taking care to set up the framework (quite literally) that we will use to record the positions in space and time of events. Care is necessary: we need to be very clear on how to measure the time of an event.

To illustrate why, let's consider a particular event: a firework exploding. The time of the explosion event is the time recorded on a clock sitting next to the firework when it explodes. This is different to the time measured by someone watching from a safe distance away, because the flash of light from the firework will take a small amount of time to reach the person watching. Light travels at approximately one foot per nanosecond, or 30.48cm in a billionth of a second, if you're of metric persuasion. I always think that, if there is a creator, this is evidence that She worked in imperial units. Another way of appreciating why we need to be careful is that we must make no assumptions regarding the rate at which all the clocks tick. We said that they are identical clocks, so you might think they all merrily tick together. But that would be an assumption, and as we will discover later on, that is wrong. This is why we have to be very clear in defining precisely how we should measure the time.

A reference frame is also a way of establishing our point of view; our perspective on the Universe. We are free to erect our imaginary frame of reference, and somebody else is free to erect their own imaginary frame. Generally speaking, any two frames of reference might be moving with respect to each other (imagine one array of clocks and rulers sliding past a second array of clocks and rulers). The range of possible reference frames is limitless, but in his Theory of Special Relativity Einstein singled out a special set of reference frames. Specifically, he introduced the idea of an 'inertial reference frame'.

You are at rest in an inertial frame if you observe that an isolated object is either sitting at rest or moving in a straight line at fixed speed. Frames that are spinning, such as the frame you are currently sitting in on the rotating Earth, are not inertial. Many of the things we take for granted in our lives, from the behaviour of storm systems to the ebb and flow of the tides, are the result of the fact that we are spinning, and therefore not in an inertial frame of reference, even though we don't feel it. We will see how this works when we explore the ocean tides and the behaviour of storm systems.

As we've already mentioned, there isn't a 'special' inertial reference frame; all inertial reference frames are as good as each other. If you're in an inertial reference frame, you are allowed to say that you are standing still, and there is absolutely no measurement you can make that will tell you otherwise. It is because we are approximately sitting in an inertial reference frame on the surface of the Earth that we don't feel as if we are moving from moment to moment.

Einstein elevated the requirement that all inertial frames are equivalent to a fundamental principle. This means that identical experiments carried out in different inertial frames will always lead to the same results. To put it another way, the laws of Nature do not change as we switch our point of view between inertial reference frames; if they did, we could tell the difference between the reference frames! I don't want to give the game away early in the chapter, but this ultimate democracy between inertial frames turns out to be such a severe constraint on the laws of Nature that Newton's laws and the laws of electricity and magnetism cannot both be right. This may not sound too serious, but we will see in Chapter Four that the laws of electricity and magnetism are one of the great pillars of physics alongside Newton's laws. They describe so many things we take for granted in our everyday lives; the action of electrical generators and motors, the formation of a rainbow, the action of lenses, the optical fibres that bring the internet into your home, and, when merged with quantum theory, the structure of atoms and molecules; the list is virtually endless. It is inconceivable that the framework we use to describe one of the four fundamental forces of Nature could be incompatible with the theoretical framework we use to describe motion. This conflict is what motivated Einstein to develop a new theory of space and time. We'll get to that. For now, let's explore the idea of describing the world from different points of view, which is to say using different reference frames, within a Newtonian framework. This will lead us to an understanding of the passing of the seasons, the rotation of storm systems and the ocean tides.

Life on an orbiting planet

The Seasons

The passage of the seasons is a gentle experience with powerful resonance. I can recite the words of hymns memorised decades ago that celebrate the great cycles of life in the North; 'In the bleak midwinter, frosty wind made moan. Earth stood hard as iron, water like a stone.' A handful of out-of-time voices drifts in the dark depths of a winter snow painted by yellowed light that falls through stained glass. 'We plough the fields and scatter the good seed on the land.' The quiet of autumn woodland in September, faded green splashed with berry red. The daily transitions are gentle, the reddening leaves and cooling of the streams subtle, but the seasonal shifts mask jarring celestial violence.

I love simple questions; they provide the opportunity to learn a lot, if not dismissed too lightly. They are also traps for the overconfident. Scientists are sometimes described as possessing a childlike quality when contemplating Nature, which I take to mean that scientists don't simply wave away questions that appear to have obvious answers without checking whether the obvious answer has content and meaning. Perhaps children have a better-developed sense of intellectual honesty. The answer to the question 'Why do the seasons pass?' has a superficial answer: 'because the Earth goes round the Sun'. But what keeps the Earth in orbit around the Sun? That also has a deceptively simple answer: gravity. But gravity is a force that acts between the Earth and the Sun, pulling them together, so why does the Earth keep orbiting and not just simply fall in? That's a deeper question.

The seasons are obviously something to do with the Earth's orbit around the Sun, which has something to do with gravity. Newton was the first to write down a mathematical model for the force of gravity.

He published it in 1687 in *The Principia Mathematica*, alongside his laws of motion. Newton's law of universal gravitation states that there is a force of attraction between all massive objects which is inversely proportional to the square of the distance between them.

$$F = G\frac{m_1 m_2}{r^2}$$

The first thing to notice is that the force of gravity acts directly along a line drawn between the centres of the Earth and the Sun, pulling them together. You may remember Newton's second law of motion from school. It is usually written as an equation:

$$F = ma$$

This says that an object will accelerate in the direction in which the force acts, and the acceleration is proportional to the strength of the force and the mass of the object. This is intuitively obvious; if you want a bus to accelerate you have to push harder than if you wanted a feather to accelerate. The force of gravity therefore accelerates the Earth directly towards the Sun. This would seem to suggest that the Sun and Earth should get closer together over time, but this doesn't happen. Why? The Earth must also obey Newton's first law of motion – the law of inertia; if no force acts on it, it will continue to move in a straight line forever. If the Sun is nearby, the force of gravity acts along a line between the centre of the Sun and the centre of the Earth. Since F = ma, this will cause the Earth to be deflected from its straight line so that it accelerates towards the Sun in the direction of the force. It will still continue happily on its way in the direction of the 'straight line', though, because no forces are acting in this direction. The Earth is therefore accelerating towards the Sun, but also flying along in a direction at right angles to the acceleration, and the net effect is that it orbits around the Sun forever. Think of the Earth falling towards the Sun but continually missing because it's also got some speed at right angles to the force that's making it fall.

There is a great deal of beautiful subtlety in the analysis of orbits. Newton discovered the family of all paths that objects will take if they move under the influence of a force proportional to the square

THE DAILY TRANSITIONS ARE GENTLE, THE REDDENING LEAVES AND COOLING OF THE STREAMS SUBTLE, BUT THE SEASONAL SHIFTS MASK JARRING CELESTIAL VIOLENCE.

of the distance between them. These curves are known as the conic sections, because they are the shapes you get if you cut through a cone at different angles (see illustration, below).

Isn't that a beautiful thing? Perhaps you can see that a circular orbit is a very special case – it only happens when the cone is sliced parallel to its base. At shallow angles the orbits are elliptical, and at steeper angles the orbits are known as parabolic or hyperbolic.

The Earth's orbit around the Sun is an ellipse. The closest approach, known as perihelion, occurs near the beginning of the calendar year around 3 January, when Earth passes within 147 million kilometres of the Sun. Six months later, our orbit carries us 5 million kilometres further out. The most distant point, known as aphelion, occurs around 3 July. The particular details of the orbit – the angle of the slice through the cone – are determined by what physicists call the initial conditions. In our description of the Earth's motion we broke things down into two parts; the Earth's straight-line motion without the Sun, and the deflection caused by the gravitational force if we put the Sun down somewhere near it. This isn't how it happened! But in this imaginary case, the initial conditions would be the initial speed of the Earth relative to the Sun, the relative positions of the Earth and Sun when we dropped the Sun in, and the mass of the Sun.

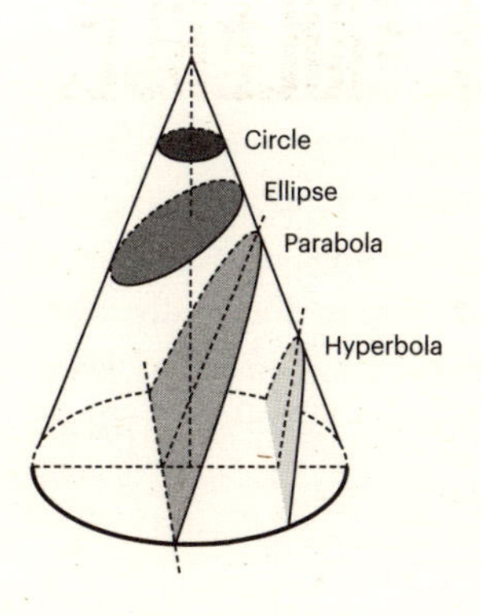

Left: A conic is a curve that is created as the intersection between a plane and right circular conic surface. The four basic conics are the circle, ellipse, parabola and hyperbola, depending on the angle of intersection.

Can you see why the details of the orbit don't involve the mass of the Earth? That's an exercise for the interested reader. All the planets move in elliptical orbits. Some comets move in parabolic or hyperbolic orbits, which means that they will only visit the inner Solar System once before escaping off into space. Halley's Comet is in an elliptical orbit, otherwise it wouldn't return every 76 years. We've built five spacecraft that are travelling on hyperbolic trajectories away from the Sun, which means that they will journey into interstellar space, never to return. They are Pioneers 10 and 11, Voyagers 1 and 2, and New Horizons. All these different paths are a consequence of Newton's law of gravitation and his laws of motion, and the particular initial conditions that started the whole thing off.

The Earth's orbit is half the explanation for the gentle passage of the seasons. To see why it isn't the whole story, consider the climate in Tasiilaq, southeastern Greenland, one of the locations we filmed in for *Forces of Nature*. Tasiilaq is a remote settlement sitting approximately 100 kilometres south of the Arctic Circle. The two thousand residents of the town experience extreme seasonal fluctuations. It's rarely what one might call warm, with summer temperatures rising to around 10 degrees Celsius on the average July afternoon. Winters, on the other hand, are brutal. The average high temperature in December is -4 degrees Celsius, and temperatures regularly approach -30 degrees Celsius. That's mild compared to northern Greenland, where a temperature of -70 degrees Celsius has been recorded. Compare that to the coldest temperature ever recorded on Earth, in Antarctica on 10 August 2010, which was -93 degrees Celsius. That's chilly.

Notice that winter is at its harshest in Tasiilaq in January when the Earth is closest to the Sun, and warmest when the Earth is furthest away. That is, of course, because the timing of winter in the northern hemisphere has nothing to do with the distance between the Earth and the Sun; it's because the Earth's spin axis is tilted at an angle of 23.5 degrees to the plane of its orbit around the Sun, as shown in the illustration opposite. In January, the North Pole and virtually all of Greenland are pointing away from the Sun and experiencing near-perpetual night. This is why it's cold. Why is the Earth's axis tilted? That's a good question.

Below: The Earth's orbit around the Sun is an ellipse, with the Sun at one focus. The Earth's spin axis is tilted at an angle of 23.5 degrees to the plane of its orbit, and it is this tilt that gives us our seasons.

Bottom: This illustration shows the current positions of four spacecraft which are leaving the Solar System on escape trajectories – our first emissaries to the stars. On this scale, the nearest star to the Sun would be approximately 100 metres away, and it would take Voyager 1 about 70,000 years to cover that distance (view from 10 degrees above ecliptic plane).

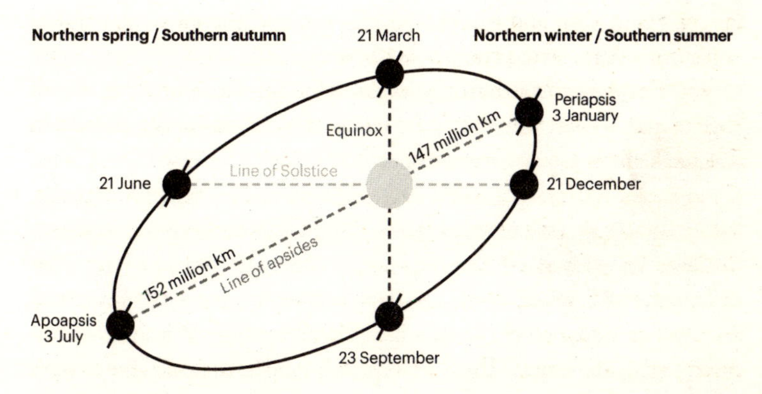

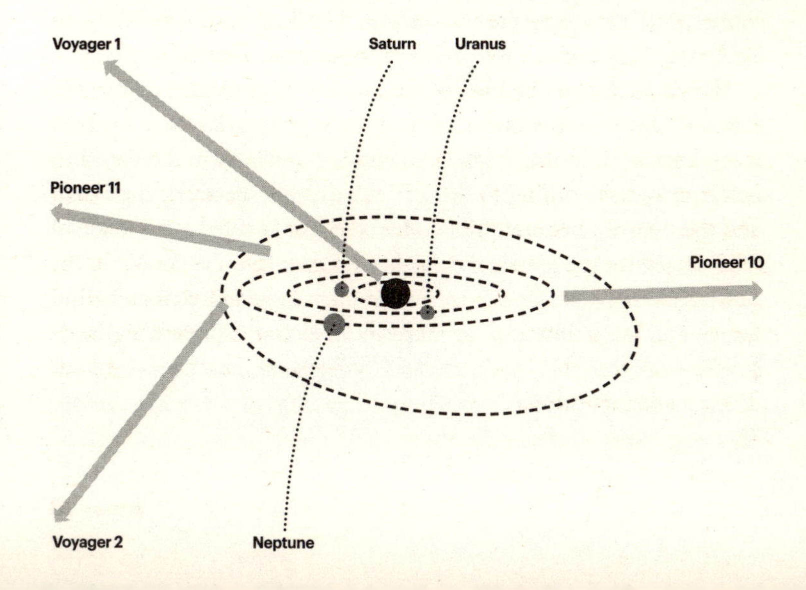

The formation of the Earth and Moon

Four and a half billion years ago, when the Earth formed, there was no Moon. Our planet was a hostile, molten ball of rock travelling around the Sun. The young Solar System was a chaotic place, with crowded orbits and frequent collisions.

Today the Earth orbits in an astronomical highway that is mainly clear of debris, which is good if you are travelling at 30 kilometres per second. A cleared orbit is one of the three definitions that the International Astronomical Union (IAU) uses to classify a planet. To clear its orbit the Earth had to go through a violent period of collisions and near-misses as smaller bodies were either thrown out of the orbit or added to the mass of the planet itself in collisions.

Not all of the objects the Earth encountered as a young planet were small. It is thought that there were dozens of proto-planets orbiting the Sun in those days, swirling around in crowded orbits, and Earth would have experienced a number of significant collisions. Direct evidence of these planetary collisions has long been erased from Earth's surface, but one particular collision left an indelible mark.

The Giant Impact Hypothesis suggests that there was a glancing collision between the newly formed Earth and a Mars-sized planet around 4.5 billion years ago, resulting in a planetary merger. The colliding planet has been named Theia, after the Greek goddess who gave birth to Selene, the goddess of the Moon. Scientists love their Greek mythology, and there is a good reason for the choice of goddess in this case. Computer simulations suggest that the collision resulted in large amounts of material from both Theia and Earth entering orbit around the battered larger planet, and over time the debris combined under the action of gravity to form the Moon. The supporting evidence for this hypothesis is strong, although, as always in science, healthy scepticism remains. Without scepticism there can be no progress. Computer simulations certainly match the details of the spins and orbit of the Earth–Moon system, but there is also physical evidence of a common origin for the system from the lunar rock samples returned by the Apollo astronauts. In particular, the abundances of oxygen isotopes ^{16}O, ^{17}O and ^{18}O in lunar rocks are near identical to those on Earth. For those who need a bit of chemistry revision, isotopes are atoms of the same chemical element but which have different numbers of neutrons in the nucleus. The most plausible reason for this similarity is that the rocks have a common origin – namely the collision 4.5 billion years ago. The Moon also has significantly less iron in its core than Earth. This is also consistent with the computer models describing such an impact. To get the spins and orbit right a glancing collision is required, and in such collisions the iron-rich cores of the colliding planets tend to merge together, leaving the iron-depleted rocks from the outer layers to form the Moon.

The Giant Impact Hypothesis is able to explain the composition of the Earth and Moon and the details of their orbits and spins. This includes the origin of Earth's tilted spin axis, angled at 23.5 degrees to the plane of the Solar System, which gives us our seasons (see page 95). I find this a wonderful thing; there are few certainties in science, but I would contend that we wouldn't be here today if our spin axis wasn't tilted. The Moon was likely formed in the event that tilted our spin axis, but in any case her presence acts to stabilise the orientation of Earth's axis, and a reasonable level of stability over geological timescales is a prerequisite for the evolution of complex life. Humans wouldn't be here without the Moon; at the very least, evolution would have taken a different path, and it is a major understatement to say that the road to humanity was convoluted. In one sense that's a superficial observation. There are a vast number of chance events in our past that could have happened differently, and changing any one of them would have meant that we wouldn't be here. We shouldn't fall into the trap of attaching particular importance to a single event; we'll leave that to the sonorous voice-overs of badly made television documentaries. The deeper unarguable point, which does bear at least a thought, is that we are very lucky indeed to be here. There cannot be any cosmic significance to our existence, because our existence is far too contingent on a series of chance events stretching back to the formation of the Solar System and beyond. Does the fact that you're lucky to be alive make you feel irrelevant or valuable? I'll leave that to you. In his essay 'Some Thoughts on the Common Toad', George Orwell reflects on the simple and available delight of noticing things like the passage of the seasons, and that is really what this book is about: 'The point is that the pleasures of spring are available to everyone and cost nothing', he writes. 'How many a time have I stood watching the toads mating, or a pair of hares having a boxing match in the young corn, and thought of all the important persons who would stop me enjoying this if they could. But luckily they can't.

'The atom bombs are piling up in the factories, the police are prowling through the cities, the lies are streaming from the loudspeakers, but the Earth is still going around the Sun, and neither the dictators nor the bureaucrats, deeply as they disapprove of the process, are able to prevent it.'

You don't need permission to do science, to think carefully and without preconception about what Nature is telling you. After all, Nature is a more reliable guide to the truth than the opinions of those incalculably lucky humans.

DOES THE FACT THAT
YOU'RE LUCKY TO BE ALIVE
MAKE YOU FEEL IRRELEVANT
OR VALUABLE?
I'LL LEAVE THAT TO YOU.

Life on an orbiting planet

Storms

The passage of the seasons is a gentle reminder that we live on a planet in orbit around the Sun. Although we're moving at close to 30km/second in orbit, we can't tell that from moment to moment because we're moving in a straight line at constant speed to a good approximation, so it feels as if we're standing still. This is why we don't feel as if we are flying through space very quickly on a ball of rock. But there is a very important caveat; we are also spinning around as the Earth rotates once a day on its axis, and this does have definite physical consequences that we experience on timescales of hours rather than months.

How do we know we're spinning?

You don't have to be particularly observant to notice that something is spinning. The Sun rises in the east and sets in the west, arching across the sky. When it sets, the stars follow suit. There is obviously something circular going on.

From the evidence available to us, we might offer two possible explanations. The first and perhaps most natural is that the Earth is stationary and the Sun and stars circle around us once a day. The other possibility is that it is we who are doing the rotating rather than the Sun and stars. Copernicus described a spinning Earth moving in orbit around a fixed Sun in *De revolutionibus*, published in 1543. He was motivated primarily by his distaste for the inelegant explanation of the observed motions of the planets against the stars laid down by the Greek astronomer Ptolemy in the second century. Observed over the course of months, the planets do not follow neat circular arcs across the sky. They perform occasional loops, reversing their motion against the

Below: This depiction of Copernicus's heliocentric system of the Universe shows the Sun, the orbits of the planets and the firmament of the fixed stars.

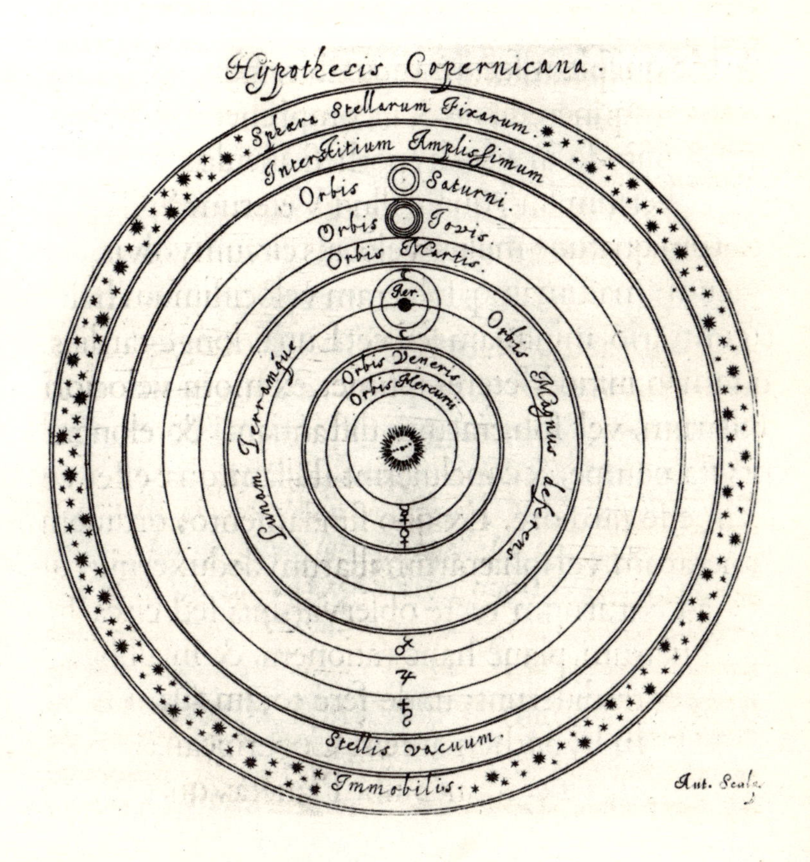

starry background. We now know this happens when the Earth overtakes a planet as it orbits the Sun. If you don't accept that the Earth is in orbit you have to come up with some other mechanism for the planetary loops, and Ptolemy's Earth-centred model, whilst delivering accurate predictions for the motions of the planets, is a terrifically messy affair. If you accept that the Earth goes around the Sun, on the other hand, you also have to come up with an explanation for day and night, which is separate from the yearly orbital motion. This is why Copernicus proposed that the Earth spins around on its axis once every 24 hours.

Copernicus's model wasn't convincing to many astronomers and natural philosophers of the day. It's revealing to read a criticism from the greatest observational astronomer of the age, Tycho Brahe: '… such a fast motion could not belong to the Earth, a body very heavy and dense and opaque, but rather belongs to the sky itself whose form and subtle and constant matter are better suited to a perpetual motion, however fast.'

Here again we see how difficult it is to accept that we live on a moving planet when we feel so powerfully that we are standing still.

Almost 150 years after Copernicus, the Italian priest and astronomer Giovanni Riccioli offered a more scientific objection to Copernicus's spinning Earth than the rather philosophical statement that it just doesn't feel right. He carried out a rather beautiful analysis of the motion of projectiles on a spinning planet in *Almagestum Novum* (New Almagest), published in 1651, when the young Isaac Newton was just 9 years old. Riccioli was concerned with laying out the evidence for and against the motion of the Earth, which he did in 77 carefully constructed arguments. Argument number 18 is an analysis of the motion of a cannonball on a spinning planet. Riccioli argued that a cannonball fired northwards (in the northern hemisphere) should follow a flight path that is distorted by the spin of the Earth. Here is what he said:

'If a ball is fired along a Meridian toward the pole (rather than toward the East or West), diurnal motion will cause the ball to be carried off [that is, the trajectory of the ball will be deflected], all things being equal: for on parallels of latitude nearer the poles, the ground moves more slowly, whereas on parallels nearer the equator, the ground moves more rapidly.'

Below: Illustration from Riccioli's 1651 *New Almagest* showing the effect a rotating Earth should have on projectiles. Riccioli's explanation for expecting a curved path is as follows: The more southerly cannon is moving faster relative to the more northerly target (E). Because the ground is moving more slowly at the target (E), it will follow a curved path and land to the right of the target at (G) instead. No such effect should be seen if the cannon is fired in the direction of the Earth's spin at an easterly target (C). This is because in this case the cannon and the target are both travelling at the same speed relative to each other and so the cannonball will fly as if the Earth is completely still. This last part of Riccioli's argument is incorrect, but he was on the right track.

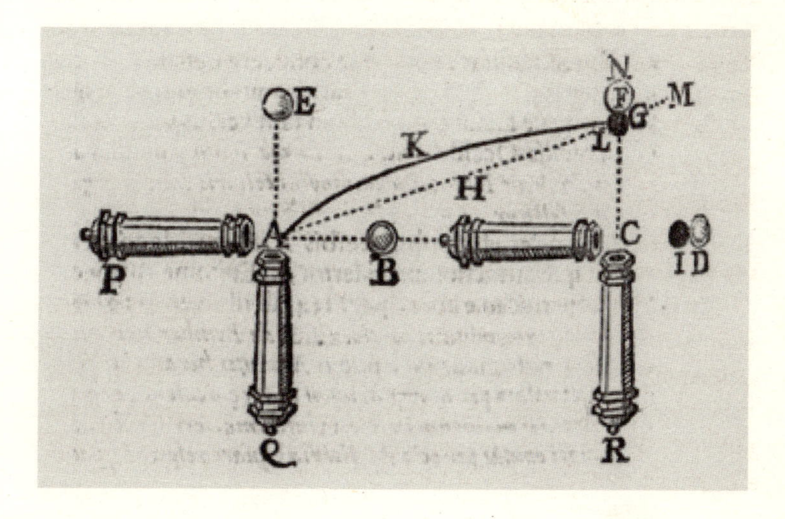

Riccioli could find no experimental evidence to show that cannonballs are deflected as they fly north, so he concluded that the Earth is not spinning. It's probably more correct to say that he reinforced his own prejudice that the Earth isn't spinning. But that's not the point. Riccioli didn't have access to good enough data to see the effect of Earth's spin on a flying cannonball, which *is* deflected in flight because of the Earth's spin. Riccioli didn't stop there, though. He also proposed that the same effect should be seen for objects falling vertically to the ground, a point he made poetically in argument number 10 of *New Almagest*:

'If an angel were to let fall a metal sphere of great weight hung to a chain, while holding the other end of the chain immobile, that chain by the force of the sphere might be extended to its full length perpendicularly toward the Earth. But following the Copernicans, it ought to curve obliquely toward the east.'

Right again, and with plenty of towers to choose from in northern Italy, Riccioli climbed to the top of the Torre degli Asinelli in Bologna and dropped some weights. He searched for a deflection in vain, which again confirmed his belief that the Earth is not spinning. His problem, again, was not his theoretical prediction (which is spot on), but the quality of his experimental data.

For those attempting to find Earthly experimental proof for the Copernican view of a Sun-centred Solar System, Riccioli's experiment was a prime target. Writing in 1679, Newton shared 'a fansy of my own about discovering the Earth's diurnal motion' with his contemporary and rival physicist, Robert Hooke. Hooke decided to attempt the experiment, culminating in a demonstration at the Royal Society on 22 January 1680. With such slight margins – a modern calculation of the deflection for an 8-metre drop is 0.3mm – the experiment failed and the records of the Royal Society give no indication that Hooke ever attempted it again.

To this day, drop-experiments such as the Torre degli Asinelli experiment proposed by Riccioli are reasonably difficult to perform, although certainly not impossible,[1] but we don't need to resort to the laboratory to observe a direct physical effect of the Earth's rotation because we have spacecraft and weather.

The grainy black-and-white image shown on page 7 of the plate section occupies an historic place in the archives of meteorology. The TIROS satellites were little spinning drums, just over a metre in diameter, and carried two wide-angled television cameras, a tape recorder for the images and a 2-watt transmitter. On 10 September 1961, TIROS-3 peered down onto the Atlantic Ocean from low Earth orbit and observed the birth of Hurricane Esther hours before its formation was spotted back on Earth.

Half a century later, the quality of space-based weather imagery is extraordinary. High-definition images allow us to keep track of the surface of the Earth and the formation of major weather systems in real time. They are ubiquitous, and because of this we know what storm systems look like. The most obvious feature is that they rotate, and the reason for this is the rotation of the Earth, as Riccioli predicted. The force that acts on weather systems causing them to rotate is the Coriolis Force, named after the French mathematician Gaspard-Gustave de Coriolis, who first published a full mathematical treatment as part of an analysis of the physics of water wheels in 1835.

Below: Trajectory of a ball rolled on a rotating disc.

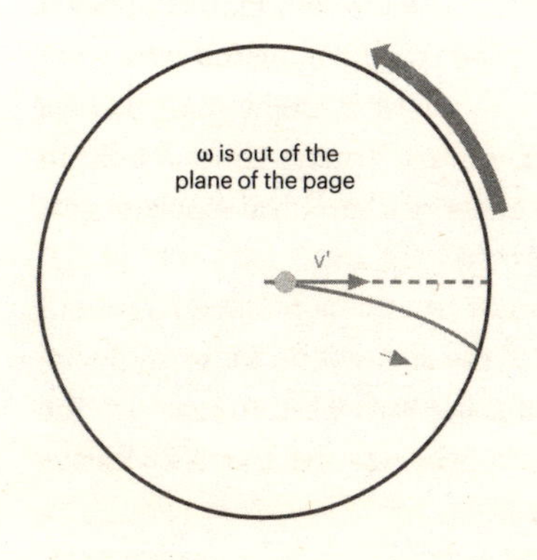

The Coriolis Force is known as a 'fictitious force', although its effects on weather systems are very real. It's called a fictitious force because it's not a fundamental force of Nature. It's not gravity, it's not electromagnetism, and it's not the strong or weak nuclear force. Rather, its origin lies in the fact that the Earth's surface is NOT an inertial reference frame. Why is the Earth not an inertial frame? Because if we stand on the surface of the Earth we are constantly changing direction as we spin around in a circle once every day. We are certainly not moving in a straight line and we are therefore not in an inertial reference frame.

How might this lead to a force 'magically' appearing? Imagine that you're sitting on a train moving at constant speed and you decide to put a cricket ball on the table in front of you. It will stay exactly where you put it. This is as it should be, because the train is an inertial reference frame and there is no experiment we can do to tell whether or not we are moving. We must all have had the experience of sitting peacefully in a train carriage, rolling gently through a station at constant speed and getting the slightly dizzying feeling that the station is drifting by. This isn't an error of perception; you are absolutely entitled to claim that you aren't moving and the platform is. If the train accelerates quickly out of the station, however, the ball will roll towards you. How should you interpret what is happening?

Newton's second law of motion states that $F = ma$. From your perspective on the train, you'll see the ball accelerate towards you on the table, and you will describe the acceleration as being due to a force acting on the ball. This force is a fictitious force. It appears *because you are no longer in an inertial frame of reference because the train is accelerating*. This might appear to be a subtle point, but it provides a way of determining experimentally whether or not you are in an inertial frame. If things in your world deviate from their state of rest or uniform motion in a straight line and the cause isn't one of the fundamental forces of Nature, then you can deduce that you are not in an inertial frame, and here is where the abstract becomes concrete. This fictitious force is very real from the point of view of the person sitting in the accelerating frame. If you were resting your face on the table when the train started accelerating, the cricket ball would hit you in the head, and there is nothing fictitious about a broken nose.

The Coriolis Force that drives the great storm systems on the surface of our planet is another very powerful example of a fictitious force.

The origin of the Coriolis Force is not as simple as the accelerating train, or for that matter as simple as Riccioli's description in his cannonball experiment; this is why it isn't called the Riccioli Force. Here is the explanation.

The Earth is a three-dimensional spherical object, which complicates things, so let's consider what happens to an object that moves around on a flat spinning disc. The arguments will be the same and easier to visualise. Imagine the rotating disc from two different perspectives. One will be that of an observer watching everything from afar – dare we say it, in an inertial frame of reference. (There is a drinking game here somewhere.) The other will be that of an observer sitting at the edge of the rotating disc, whizzing around with it. This is our situation as we sit on the surface of our spinning planet.

Now imagine that the rotating observer decides to throw a ball directly towards the centre of the disc. From their perspective, the ball sets off happily in the direction in which it is thrown but immediately starts to curve away in the direction of rotation. What is happening? It's easiest to see from the perspective of the observer watching from afar (see page 105).

From the distant perspective, the ball is flying around in a circle with the disc, before it is thrown inwards. When it's thrown, it hangs on to the initial speed it had in the direction of rotation. This is the law of inertia again. Nobody pushed on the ball in the direction of the spin of the disc, which is known as the tangential direction, so it simply keeps on going. As it rolls inwards, however, it finds itself travelling too fast in the tangential direction for the inner parts of the disc. This is because the points closer to the centre have less far to travel to circle once around, so they must be travelling more slowly than the points further out. As a result, the ball gets ahead of the disc and curves away in the direction of rotation. From the distant observer's perspective, there is no force acting on the ball. The curved path is explained purely in terms of the rotation of the disc.

From the rotating observer's perspective, however, there appears to be a force acting on the ball in accord with Newton's first law, because it doesn't travel in a straight line relative to them. This is the

Coriolis Force. It acts at right angles to the direction of motion of the ball, deflecting it onto a curved path. On the surface of the Earth, the Coriolis Force always pushes objects moving in the northern hemisphere to the right, and objects in the southern hemisphere to the left, if we view the Earth as being orientated with the North Pole at the top. At the Equator, the Coriolis Force pushes neither to the right nor the left, although it does try to lift an object gently off the surface! Such is the complexity of a rotating three-dimensional sphere rather than a disc, but the principle is the same.

We can now see why storm systems rotate the way they do on the surface of the Earth. Large bodies of air do not move in straight lines because of the action of the Coriolis Force. A cyclone is a region of low pressure. The higher-pressure air around it will fall inwards to try to equalise the pressure. In the northern hemisphere, the moving air will experience a Coriolis Force to the right as viewed from above, and therefore will rotate in an anti-clockwise direction around the low-pressure area. In the southern hemisphere, a cyclone will rotate in a clockwise direction because the inward-falling air is deflected to the left. This is why the hurricanes that form every year in the Atlantic which threaten the Caribbean Islands and the southeastern states of America always rotate anti-clockwise, whereas the tropical cyclones (the name for a hurricane that forms in the southern hemisphere) that batter the Pacific Islands are always rotating in the opposite direction.

For anti-cyclones, the opposite is true. The air flows outwards from a high-pressure central region, and the deflection to the right by the Coriolis Force in the northern hemisphere induces a clockwise rotation.

As well as creating the distinctive spirals of storm systems as seen from space, the Coriolis Force also increases the strength of the storms. The stronger the deflection of the air current around a high-pressure system, the faster it will rotate. This is one reason why the most powerful storms in the Solar System occur on faster-spinning planets. Jupiter is not only the most massive planet, it is also the fastest rotating, spinning once on its axis approximately every 9.8 hours. The most recognisable storm system in the Solar System is the Great Red Spot, a spiralling storm that has raged on the gas giant for at least two hundred years, but probably far longer. Famously large enough to swallow the Earth whole, it is 20,000 kilometres long, 12,000 kilometres wide and boasts wind speeds of up to 700km/hr. The Coriolis Force generated by the size and rotation speed of Jupiter is a significant contributing factor to the power and size of the Great Red Spot and the many other storm systems that rage through Jupiter's swirling clouds. The Great Red Spot is an anti-cyclonic (high-pressure) storm in Jupiter's southern hemisphere and, just as here on Earth, it therefore rotates in an anti-clockwise direction. The laws of Nature are universal.

The reason for the rotating storms on Earth and across the Solar System is interesting in itself, but there is a deeper reason why we've spent time studying the Coriolis Force. It appears in our description of the physics when we try to explain a real-world natural phenomenon from different perspectives – that is to say from different frames of reference. Hold that thought, because we'll come back to it.

Recall that we began this chapter musing about the nature of space and time, and hinting at the rather wonderful suggestion that events in our past may have an existence beyond our memories. This chapter is a wandering adventure in a sense; our explanations of natural phenomena will serve to illustrate something we need to know on the road to relativity. Let us explain one more everyday physical phenomenon that requires us to jump between different frames of reference to understand: a classic problem in physics – the ocean tides.

[1] For a height of 97 metres (the height of the Asinelli tower) at a latitude of 44.5 degrees north (Bologna) and with $\omega = 7.3 \times 10^{-5}$ /s (the angular speed of Earth), the deflection is equal to 1.8 centimetres. The details of this calculation can be found, for example, in Forshaw and Smith, *Dynamics and Relativity* (Wiley).

Life on an orbiting, spinning planet

The Tides

The ebb and flow of the tides creates a dramatic, recurring and rapid transformation of Earth's coastline. With a little patience and a comfortable deckchair you can watch the landscape change before your eyes. Geological in timescale it isn't. The Bay of Fundy on Canada's east coast holds the record for the greatest tidal range: 56 feet, as measured by the Canadian Hydrographic Service at Burntcoat Head. I am delighted to leave this measurement in feet as a celebration of cultural diversity.

The origin of the tides is an ancient puzzle. The connection between the tides and the lunar cycle has been known for well over 2000 years, but the recognition of patterns and the prediction of high, low and spring tides does not require an understanding of the underlying mechanism. If all you want to do is sail, you don't need to know why; you just need to know when. With the emergence of a heliocentric model of the Solar System in the sixteenth century, the understanding of the origin of the tides received a great deal of attention from the astronomers of the day because it presented an Earth-bound phenomenon that appeared to be connected to the motion of Earth, Moon and Sun. Johannes Kepler asserted that the tides were created by a force of attraction exerted by the Moon on the Earth's oceans, but was unable to provide a mechanism to explain the force. Galileo disagreed, and in an increasingly fractious dialogue proposed the counter-argument that the tides are a result of the Earth's rotation and revolution around the Sun: 'Among all the great men who have philosophised about this remarkable effect, I am more astonished at Kepler than at any other. Despite his open and acute mind, and though he has at his fingertips the motions attributed to the Earth, he nevertheless lent his ear and his assent to the Moon's dominion over the waters, to occult properties, and to such puerilities.'

'Puerilities' is a word I intend to use more often. In an essay written in 1616 entitled 'Discourse on the Tides', Galileo likened the movement of the Earth's oceans to the movement of water in a vase. He reasoned that because the water is distorted by changes in the orientation and acceleration of the vase, so the oceans are distorted in their movement by the orientation and acceleration of the Earth. He posited a mechanism of positive and negative acceleration to explain the back-and-forth motion of the tides, a theory that has often been labelled his 'great mistake'. The irony is, both Galileo and Kepler were partly right. Here is the explanation for the origin of the tides.

Kepler was correct in the sense that the tides are caused by the Moon's gravitational effect on the Earth. He didn't put it in those terms, of course, because Newton had yet to publish his theory of Universal Gravitation. Galileo was correct because the Earth is accelerating. He just didn't appreciate towards what.

Let's accept that the tides have something to do with the Moon; its orbit can be described in precisely the same way that we described the Earth's orbit around the Sun. The Moon is being pulled towards the Earth by the force of gravity but is continually missing it because it continues to try to move in a straight line, in accord with the principle of inertia.

We now need to introduce Newton's third and last law of motion. It states:

To every action, there is an equal and opposite reaction.

This means that forces always come in pairs. If the Earth exerts a gravitational pull on the Moon, the Moon exerts an equal and opposite gravitational pull on the Earth. This means that the Earth has to fall towards the Moon, accelerated by the force of gravity along a line connecting their centres. Why doesn't the Earth career towards the Moon? For the same reason that the Moon doesn't career towards the Earth – because it falls and misses. The Earth must also be in orbit! But around what? The answer is that we were a little lax in our language when we said that the Moon orbits around the Earth. It does to a good approximation, but in fact it orbits around a point slightly displaced from the centre of the Earth known as the centre of mass of the Earth-Moon system. To get an instinct for what's happening, imagine two moons of equal mass orbiting around each other in circular orbits. Everything is perfectly symmetric, and they both orbit around a point that is equidistant between their centres. This is called the centre of mass of the system. If one of the moons is more massive than the other, the centre of mass will be closer to the massive moon, and they will both orbit around this offset point. The Earth is 81 times more massive than the Moon, so the centre of mass point about which they orbit is very close to the centre of the Earth, but not quite at the centre; it is displaced by 4700 kilometres, which is about 1/81 of the distance between the Earth and the Moon. This is why it's superficially reasonable, but not accurate, to say that the Moon orbits around the Earth. It's only reasonable in a superficial sense because it dodges the problem of how the Earth can accelerate towards the Moon – as it must – and keep missing!

The fact that the Earth is in orbit around the centre of mass of the Earth-Moon system is critical to an understanding of the tides. The key is to switch perspective, or frame of reference, just as we did when we explored the Coriolis Force and its effect on storm systems. We're hopping between reference frames again, searching for fictitious forces – physicists are always doing this because it's bloody useful, and we know how to do it! (See illustration below.)

Let's picture what's happening from the point of view of an observer sitting at the centre of the Earth. This is the point that is orbiting around the centre of mass of the Earth-Moon system. We

Below: The centre of mass.

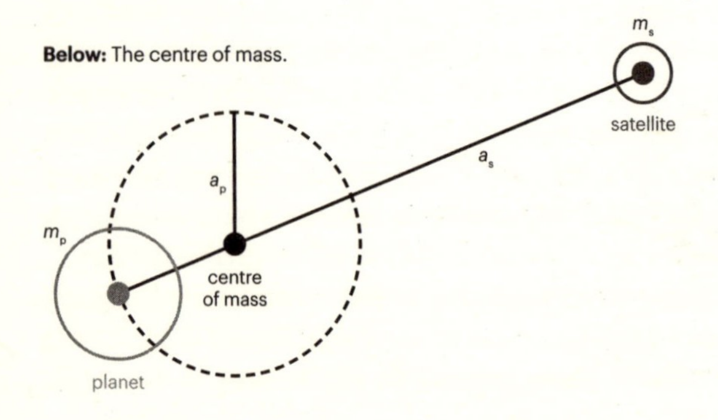

can assume that the centre of the Earth is going round in a perfect circle, which it very nearly is. This reference frame is not inertial, because it's rotating, and we will therefore expect fictitious forces to be present. But which? This time it's not the Coriolis Force, which appears when things roll around in rotating reference frames, but the Centrifugal Force. What does this one do? From the point of view of the observer sitting at the centre of the Earth, the force of gravity is accelerating the Earth towards the centre of the Moon in a straight line, in accord with Newton's law of universal gravitation. And yet, the Earth doesn't approach the Moon – it stays a fixed distance away from the centre of mass of the Earth-Moon system if the orbit is circular. This must mean that the observer at the centre of the Earth experiences a force acting with the same strength as the Moon's gravitational pull, but in the opposite direction, to precisely cancel it out. This force, equal and opposite to the gravitational force, is called the Centrifugal Force. It's none other than the familiar force we experience if we sit on a fast-rotating fairground ride. We are thrown outwards, and if it's the right sort of ride, the little cars we sit in will rise up and outwards as the speed increases. The force that does this is the Centrifugal Force.

Great. But what's that got to do with the tides? We're about halfway through, so perhaps you should have a break for a cup of tea and come back refreshed. As an aside, I find something amusing about this explanation for the tides, which is quite a wonderful

Below: The tides.

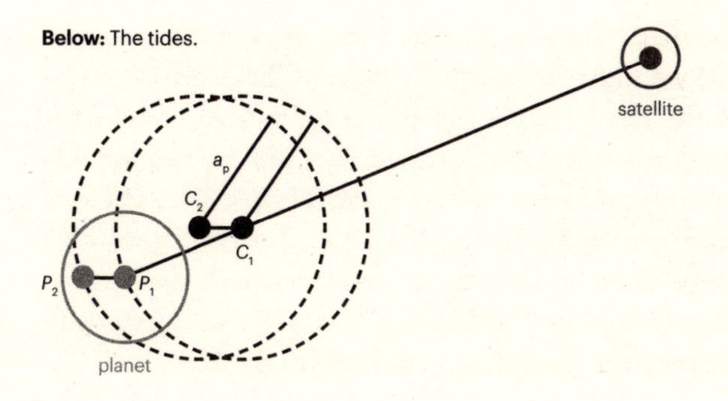

explanation if you have the patience to follow it. Let me tell you what I find amusing. Imagine, as I tell you, that there is a hint of Joe Pesci in *Goodfellas* in my voice. I have a love–hate relationship with television. I love most of it, to be honest, but I sometimes find it a superficial medium. The trick is to find a way of exploring ideas in sufficient depth within a television programme that is the length of a single undergraduate lecture, in a visual and entertaining way. I get into a lot of 'creative debates' about the definition of 'sufficient depth', as you might imagine. Usually we are exploring grand ideas about the origin of the Universe or the beginning of life on Earth, and because the answers to these ideas are speculative, there is room for a bit of hand-waving. In a programme about tides, however, there can be no hand-waving because the reason for the tides is known. I think the tides are a good thing to explain. But I offer a wry smile. 'Sufficient depth' is a well-defined concept in this instance. It is defined as being *the* explanation. Such is the trap, lying in wait for the unwary television executive, in wanting to make a television series about simple questions that actually have answers, rather than complicated questions that don't.

Here is the rest of the explanation for the tides. Recall that the Earth is in a little orbit around the centre of mass of the Earth-Moon system. At the centre of the Earth, the Moon's gravitational pull is perfectly balanced by a fictitious force called the Centrifugal Force. Now consider a point on the surface of the Earth directly beneath the Moon. That too will be in a little orbit, and it will have to go around in a circle of precisely the same radius as the point at the centre of the Earth, because the Earth is a solid ball of rock and a point on the surface can't move in a different way to the centre. This means that the Centrifugal Force experienced at a point on the surface beneath the Moon must be exactly the same as that experienced by the centre of the Earth. But – and this is the crucial point – the Moon's gravitational pull at the Earth's surface directly beneath it is stronger than it is at the centre of the Earth, because the surface of the Earth is closer to the Moon than the centre. The two forces won't precisely balance! There will be a little too much gravitational pull at the surface directly beneath the Moon, and it is this little extra pull that deforms the oceans and raises a tide beneath the Moon.

Now consider the situation on the other side of the Earth. Again, the Centrifugal Force must be the same, because every point on the Earth's surface has to orbit in a circle of precisely the same radius as every other point, but now we are further away from the Moon than the centre of the Earth is, so we'll experience a weaker gravitational pull from the Moon. This means that the Centrifugal Force, which always points away from the Moon, will be slightly too large, and this will also result in oceans being deformed away from the surface, raising a tide. This is why there are two tides on Earth every day – one beneath the Moon and one on the opposite side of the planet.

The tidal forces are the result of the imbalance between the Moon's gravitational pull and the Centrifugal Force, which is present because the Earth is orbiting around the centre of mass of the Earth-Moon system. Although we usually perceive them because of the large deformation of the surface of the oceans, they are sufficiently large that the Earth's crust is deformed by a measurable amount, shifting the rocks every day by as much as half a metre. This is not a great shift, but GPS systems are adjusted to take account of the changes in the Earth's gravitational field caused by the rock tides, and geologists monitor the impact of these tides on the Earth's fault lines and the potential they have to trigger earthquakes and volcanic eruptions.

Our explanations of rotating storm systems and tides are quite beautiful in my view, because they embody one of the central themes of this book – that apparently complex and disconnected naturally occurring phenomena can be explained using a simple, underlying framework – in this case Newton's laws of gravitation and motion. I don't believe we need a reason to seek an explanation for these things beyond the fact that it's interesting and fun. But the explanations of the tides and the rotation of storm systems both benefited from us jumping between different frames of reference, which is to say looking at physical phenomena from different points of view. This idea is the launch-pad for something deeper. As we discussed at the beginning of this chapter, Albert Einstein elevated the idea that the laws of Nature *must* take the same form in all frames of reference to a fundamental principle. Our Universe is built this way. Implementing this requirement forced him to discard Newton's laws and redraw our intuitive picture of space and time, the grand arena that is so very tempting to take for granted.

Einstein's Theory of Special Relativity

The subject of motion is unexpectedly rich. Subtleties are evident even in Newton's *Principia*. After dealing with the motion of objects in general, and developing many of the tools that modern-day physicists take for granted, Newton's focus turned to the motion of the planets around the Sun in order to address age-old questions about the tides and the passing of the days, months and years. He was also keenly aware that an understanding of space and time is necessary, and he carefully stated his assumptions about the existence of absolute space and absolute time. That Newton felt it necessary to state that absolute time exists *as an assumption* is, to my mind, a clear example of his brilliance as a physicist. Newton treated this assumption as we now treat the law of inertia; as an axiom, in agreement with observations at the time, but not provable from first principles. It is a remarkable thing that he identified such an assumption and considered it worthy of note, even though in the seventeenth century, and surely today in most people's minds, it must 'go without saying' that there is not much to say about time other than that it is absolute and that it ticks. And so we return to our musings at the beginning of the chapter about Monet's field of poppies, vanished forever – perhaps – with the passing of the years. We are now in a position to explore the tantalising 'perhaps'.

Why did Einstein replace Newton's laws of motion?

Central to our exploration of motion has been the idea of an inertial frame of reference. If you've grown weary of the term, if you recall I suggested a drinking game. If you go down this route, you are about to discover a link between vintage wine and memories.

To recap, the idea is that it isn't possible to work out which inertial reference frame you are in; they are all absolutely equivalent to each other and the notion of 'at rest' is always a relative one. In simpler language, this means that you can't tell whether or not you are moving. If you accelerate, the story is different, and fictitious forces appear. Albert Einstein thought very deeply about these ideas – more deeply, in my opinion, than anyone else. Einstein is the archetypal wild-haired, sockless genius. In later life he looked otherworldly, appearing to inhabit an abstract space beyond Earthly trivia alongside his theories. This is, of course, a cliché; Einstein was a great physicist, but he discovered no Royal Road to understanding because no such road exists. He worked hard, thought deeply and learnt how to do sums. That said, his theories of relativity are certainly amongst the greatest of human achievements. Over a century after their publication, they are still part of the essential foundations of modern physics.

Einstein discovered his Theory of Special Relativity by elevating the idea that all inertial reference frames are equivalent to a great principle; an axiom; a fundamental property of our Universe. It was his guiding light. To understand why this was so important to Einstein, we need to revisit a concept we explored in Chapter One, symmetry.

The statement that all inertial frames are equivalent is a statement of symmetry. If you recall, symmetry in mathematics and physics means doing something with the result that nothing changes. A square has a particular symmetry in the sense that we can change our point of view by rotating around the square by 90 degrees and everything will look the same. We can ask a similar question about physical laws such as Newton's laws of motion. Do the laws remain the same if we change our perspective? One such symmetry relates to the question: do the laws of Nature look the same in all inertial frames?

Here's another way of looking at it. The laws of physics describe real things and how they behave. Newton's laws, as we've seen, say that a rolling ball will continue to roll in a straight line unless acted upon by a force. If there is a force, the equation that describes what will happen to the ball is $F = ma$. Let's imagine that we are watching a rolling ball, and we decide to change our perspective by hopping into a different inertial reference frame. We will choose a frame of reference that is flying towards the rolling ball and see how our description of what's happening changes. Note well that 'our description of what's happening' is another way of saying 'the laws of Nature', so what we're really saying is that we want to know how the laws of Nature change. The ball will still appear to move in a straight line because there are no forces acting, but its speed will look different. If we fly towards the ball at 20 m/s, and the ball was rolling towards us at 10 m/s, then common sense informs us that we'll see the ball rolling towards us at 30 m/s. As long as we account for the change in speed by adding the speed up in this way, we can use Newton's laws and we'll get all of our predictions correct. Our description of the physics of the situation is left unchanged by our shift in perspective. This is a symmetry of Newton's laws; they remain the same if we jump between inertial frames of reference and keep track of the change in the speeds of all the objects in a simple and intuitive way.

A little piece of jargon: accounting for the change in speed in this way is known as a Galilean Transformation, in honour of Galileo. In full physics mode, we can say that Newton's laws are invariant under Galilean Transformations – this is a symmetry of Newton's laws.

Now let's think about our explanations of storms and tides. These involve situations in which Newton's laws are not the same when we hop into a different frame of reference. In a rotating frame, a rolling ball curves and doesn't continue in a straight line due to the appearance of the Coriolis Force. We account for this by changing Newton's second law in the rotating reference frame. It doesn't look like $F = ma$ any more. It changes into $F + F_{cor} = ma$, where F_{cor} is the Coriolis Force. Similarly, when we think about the origin of the tides, we jumped into the rotating reference frame of the Earth orbiting around the centre of mass of the Earth-Moon system, and saw that $F = ma$ changes into $F + F_{cen} = ma$, where F_{cen} is the Centrifugal Force. In both cases, Newton's laws do *not* look the same in the rotating frames because extra 'fictitious forces' appear. We can say that Newton's laws are not invariant when we transform from an inertial reference frame into a rotating reference frame. Incidentally, the Centrifugal Force and the Coriolis Force are always both present, but for the tides the Coriolis Force isn't important, whilst for cyclones and anti-cyclones, the Centrifugal Force isn't important.

We've taken quite a bit of time to discuss these ideas because they are absolutely central to modern physics – and to understanding why Einstein wrote down his theory of relativity.

Einstein was the first to take a very important and, at first sight, rather odd fact seriously. Unlike Newton's laws, the laws of electricity and magnetism are not invariant under Galilean Transformations. They do not look the same in all inertial frames of reference if you change all the speeds in the way you do for Newton's laws to account for the shift in perspective. This means that Newton's laws and the laws of electricity and magnetism are not consistent with each other! This was the situation that Einstein faced in 1905.

The reason why the laws of electricity and magnetism cause a problem is a simple one, but a little bit of history is in order first. During the nineteenth century, the exploration of electricity and magnetism was at the cutting edge of physics. The names of many of the scientists are remembered in the language we use to speak about electricity today: André-Marie Ampère gives his name to the Amp, the unit of electric current, and the Volt is named after Alessandro Volta. The greatest experimental breakthrough came during 1831 and 1832 when, in a series of experiments at the Royal Institution and Royal Society in London, Michael Faraday discovered electromagnetic induction, and in doing so invented the electric generator and laid the foundations for the modern world.

During the 1860s, the Scottish physicist James Clerk Maxwell discovered a unified theoretical description of all electrical and magnetic phenomena. Maxwell's equations are one of the great achievements of the human mind. Einstein later described Maxwell's work as 'the most profound and the most fruitful that physics has experienced since the time of Newton'. The equations are so beautiful that I can't resist showing them to you. To hell with those who think equations reduce the number of sales of popular science books. Here they are:

$$\nabla \cdot \mathrm{E} = 0 \qquad \nabla \times \mathrm{E} = -\frac{\partial \mathrm{B}}{\partial t},$$

$$\nabla \cdot \mathrm{B} = 0 \qquad \nabla \times \mathrm{B} = \frac{1}{c^2}\frac{\partial \mathrm{E}}{\partial t}.$$

The Es and Bs stand for electric and magnetic fields, the basic building blocks of Maxwell's description of electric and magnetic phenomena. Written in this notation, there are only two other letters in the equations: t stands for time and c stands for the speed of light. This is the key that unlocked the door for Einstein. The speed of light enters Maxwell's equations as a constant – a fundamental number that does not change. It is one of the axioms – the building blocks of our Universe. It is a speed upon which everyone agrees, irrespective of which frame of reference they are in. This is shocking, and looks like a disaster for physics. How can it make any sense that everyone agrees on the speed of light, irrespective of what frame of reference they are in? Recall our example of jumping between different reference frames and observing a rolling ball. All we had to do was add all the speeds together in the intuitive way encoded into the Galilean Transformations and all is well. Maxwell demolishes this idea.

Imagine someone holding a torch. Light streams out of the torch at the speed of light: 299,792,458 metres per second. Now imagine someone else looks at the situation from a different inertial frame of reference, flying towards the torch at half the speed of light. We might expect that we will be able to describe everything in either frame as long as we add the speeds, in accord with the Galilean Transformations. The person flying towards the torch would conclude that the light whizzes past them at 450,000,000 metres per second – which is c + ½ c. Maxwell's equations demand that this is not the case. They say that both observers measure the speed of light to be precisely equal to 299,792,458 metres per second. The speed of light doesn't change, irrespective of how you look at it. It is a constant – a fundamental property of Nature.

If this sounds weird, it is. I have no way of explaining why, other than to say that our Universe is constructed like this. Maxwell's equations are correct. The statement that the speed of light is a constant in *all inertial frames of reference* is on the same footing as the principle of inertia. It is because it is.

Einstein's brilliance – let us call it genius – was to take Maxwell's equations at face value and insist that when we hop between inertial frames of reference we keep the speed of light the same. We are not allowed to add velocities in the way that we have been doing; it is simply wrong. The Galilean Transformations are wrong, and therefore Newton's laws, which possess the symmetry represented by the Galilean Transformations, are also wrong.

Somewhere in spacetime

We can now bring all these ideas together. Einstein rebuilt physics from the ground up by insisting on two axioms, which are known as Einstein's postulates. The first is one with which we are very familiar indeed.

The laws of physics are the same in all inertial frames of reference.

The second postulate is the one that comes from taking Maxwell's equations at face value:

The speed of light in a vacuum is the same in all inertial frames of reference.

If we were writing a physics textbook, we'd now proceed to derive all the consequences of these two postulates, and in the process discover treasures such as $E = mc^2$ – the statement that mass and energy are interchangeable. This isn't a textbook. Here, we want to explore a very particular consequence of Einstein's two postulates: the idea that space and time are not what they seem.

Let's return to the beginning; the moment at which Monet sat down in a field of poppies just outside Argenteuil and, lungs filled with the scents of a late-spring afternoon, dabbed a delicate spot of red paint onto his canvas. The position in space and time of the dab of red paint is known in the language of relativity as an *event*. Because we live in three-dimensional space, we need three numbers to describe the position of the painted poppy on the canvas. These numbers could be the latitude and longitude of the easel in the poppy field and the height of the canvas above sea level. These three numbers specify *where* the event happened. We also need a time and date to specify *when* it happened; noon on 26 May 1873. An event in space and time has four co-ordinates; three to specify its position in space, and one to specify its position in time.

Now consider another event. As the light fades, Monet slips the half-finished canvas under his arm, walks back to his room in the village and closes his door. The click of the lock marks another event, with a different latitude, longitude and height above sea level and a different time, by his watch. It's now 8pm on 26 May 1873.

Let's imagine that Monet decided to measure the distance between his easel and his door and found it to be precisely 2 kilometres, and that they are at the same height above sea level. This is the distance in space between the two events. The difference in time is 8 hours, by Monet's watch.

Newton, and everyone else before Einstein came along, would agree with the common-sense notion that any observer who decided to measure the distance between Monet's easel and door and the time between the dab of paint and the click of the lock would be in complete agreement with Monet, assuming that their rulers and watches were accurate and synchronised. Einstein discovered that, if he imposed his two postulates, this is not the case. Different observers do not agree on the spatial distance and temporal difference between events. Let's be specific. Imagine that an enterprising French lady with access to a futuristic aircraft was flying past Monet on 26 May 1873 at half the speed of light. She would measure the time difference between Monet's dot on the canvas and the click of his lock to be 9 hours and 14 minutes and the distance between the easel in the poppy field and his door to be 1.73 kilometres. This discrepancy has nothing to do with the way time and distance are measured, or the measuring devices used. Furthermore, neither Monet nor the aviator is wrong; each is absolutely entitled to claim that their measurements are correct. Rather, Einstein discovered that in reality there is no such thing as absolute time and no such thing as absolute space. Let's repeat this, because it's very odd. From the point of view of the aviator, Monet's time passes more slowly than hers, which means that Monet ages more slowly than she does, and Monet really does walk 1.73 kilometres on his way home. The converse is also true. If Monet glanced up and saw the aircraft fly by, he would see the aviator's clock ticking more slowly than his, and he would conclude that she was ageing more slowly than he. He would also conclude that her aircraft is 0.866 times shorter than it appears to her. Arguably he wouldn't have continued to paint a poppy field had this really happened, but the point is that this is not theoretical; the effect is real. Nature really is constructed this way. The slowing down of moving clocks is known as time dilation, and the shrinking of moving objects is called Lorentz Contraction.

If you are comfortable with a bit of mathematics, you'll find a derivation of the result that Monet's clock runs slow as viewed by the aviator, and by how much, on page 137. The conclusion follows directly from Einstein's two postulates, and the argument is quite simple and requires no mathematics beyond Pythagoras's theorem. If you're happy to take our word for it without reference to page 137, then accept it and read on!

The reason why Einstein's theory predicts that distances in space and intervals of time are not the same in different frames of reference are his two postulates – the requirement that the laws of Nature take the same form in all inertial frames of reference and that the speed of light is constant in all inertial frames of reference. These two postulates imply that moving clocks run slow, as proved on page 137. In more precise language, Einstein had to replace the Galilean Transformations, which tell us how to switch between different inertial frames, with a new set of equations called the Lorentz Transformations. Lorentz Transformations leave the speed of light the same, as required by the second postulate, but there is an apparently terrible price to pay: Distances in space and intervals in time do change under Lorentz Transformations: moving rulers shrink and moving clocks run slow!

So where does this leave us? We've discovered that space and time are not as they seem, if we accept that the speed of light must remain constant for all observers. There is no such thing as absolute space, because observers moving at different speeds relative to each other disagree on the distance between events. The comfortable picture of the Universe as a big box, where every star, planet and galaxy has a well-defined place, cannot be right, because the distances between the stars, planets and galaxies cannot be defined in a unique way. Similarly, there is no such thing as absolute time, because it is not possible to define the time between events in a unique way.

This is fun, and strange, but it also presents a serious problem for physics. The problem lies in Einstein's first postulate: *the laws of physics are the same in all inertial frames of reference.* The laws of physics are the tools that we use to predict the outcome of real-world experiments; they are descriptions of Nature. If they are to be the same in all inertial frames, then it follows that they should be constructed out of quantities that are the same in all inertial frames. But the laws of physics we learn at school concern distances measured by rulers and times measured on clocks. Think about Newton's second law of motion, $F = ma$, which describes how fast an object of mass, m, accelerates in response to a force, F. Acceleration is measured in metres per second squared – a quantity that involves changes in distance over some time interval. But since we've discovered that distance intervals and time intervals are not the same in all reference frames, it follows that Newton's laws are not the same either! This looks like a disaster.

It isn't, fortunately, because Einstein found a way out. He discovered that, whilst the distance in space between two events and the difference in time between two events each change, there is a quantity that does *not* change if we switch perspective between inertial frames: the distance in space *and* time, taken together in a very special way.

If we call the distance between Monet's easel and door Δx and the time difference between the dab of red paint and the click of the lock Δt, then the 'distance' $\Delta s^2 = c^2\Delta t^2 - \Delta x^2$ does not change. Both the aviator and Monet agree on Δs, even though they disagree on Δt and Δx. The quantity Δs is known as the distance in spacetime between the two events. The speed of light, c, has entered the equation in a rather subtle way, multiplying the time difference Δt. Why? One thing we can say immediately is that some speed or other had to be there to make the definition of the distance in spacetime sensible. Let's say we chose to measure time differences Δt in seconds and distances between events Δx in metres. We can't simply subtract something in seconds from something in metres – that's like subtracting five apples from ten oranges. But if we multiply the time difference in seconds by a speed, which is measured in metres divided by seconds, then we get the object c Δt, which is measured in metres, and we can happily go ahead and subtract Δx from it. That argument doesn't inform us what value c should take, but it does tell us that it has to be some speed or other.

In an undergraduate lecture course on physics, we would now proceed to consider how energy and momentum are treated in special relativity and show that this special speed can be interpreted as the speed of massless particles. Coincidently, as far as we know, photons happen to be massless and therefore travel at the special speed c – and this is why we call it the speed of light.

The fact that distances between events in spacetime are agreed upon by everyone suggests that we should rebuild our laws of Nature out of quantities like Δs, and this is precisely what Einstein did, replacing Newton's laws and quantities familiar to physicists such as energy and momentum with spacetime versions. This is where $E = mc^2$ comes from. I think it's quite satisfying that Nature is constructed in this way. Events, after all, form the narrative of our lives. We don't separate our memories into separate spatial and temporal components. I remember a perfect summer's day in August 1972 when a yellow sun lifted the scents of the lawn and Doppler-shifted bees drowned out the hum of the town. We set up a paddling pool in my parents' garden and played in the water so long we chafed our thighs. I remember this as an event, not a moment with a separate latitude, longitude and time stamp.

There is a vivid way of visualising these ideas known as a spacetime diagram. In order to draw it, we can represent the position of events in space along the horizontal axis and the position of events in time on the vertical axis, as shown in the diagram on page 128. We've neglected two spatial dimensions here for clarity because they don't matter for our argument, and it's hard to draw a four-dimensional diagram on a piece of paper. Let's draw my life as a spacetime diagram. It's important to define precisely what frame of reference we're in when we draw a spacetime diagram. In this case, Oldham Royal Infirmary, where I was born on 3 March 1968, will be our frame of reference (which we'll assume to be an inertial frame). This means that we set up a grid of Oldham rulers and Oldham stopwatches at rest relative to OldhamRoyal Infirmary. We agree to zero the Oldham stopwatches at the moment of my birth, and because I was born inside Oldham Royal Infirmary, the co-ordinates of my birth event are x = 0, ct = 0, where x is the distance from Oldham Royal Infirmary and t is the time as measured by the Oldham stopwatch at position x = 0. We'll label this event '3 March 1968', and it sits at the origin of the spacetime diagram.

We can now add some more events. In August 1972, I was four kilometres away from Oldham Royal Infirmary in my paddling pool. The time on the Oldham watch at that point reads 4½ years. On 3 March 1989 I was in Florence, Italy, on a tour bus after playing a show with my band Dare while supporting the Swedish rock band, Europe. I know. That's 21 years as measured on the Oldham watch, and I'm around 2000 kilometres from Oldham Royal Infirmary. One more. On 2 September 2009 I was in one of my favourite countries, Ethiopia, filming for *Wonders of the Solar System* at the Erta Ale lava lake with my friendly guard from the Afar tribe.

I could mark every event in my life this way, as measured by the time on the Oldham watches and the distance from Oldham Royal Infirmary. The resulting line on the spacetime diagram is called my worldline. It represents every moment in my life at the locations measured by the Oldham watches and the Oldham rulers. Remember Hermann Weyl's evocative quote: 'Only to the gaze of my consciousness, crawling along the lifeline of my body, does a section of this world come to life as a fleeting image in space which continuously changes in time.' This is what he meant.

There is another feature of the spacetime diagram that we must mention; the diagonal lines passing through 3 March 1968. These are known as a lightcone, and lightcones play a very important role in relativity. To understand what they are, imagine that someone decides to flash a beam of light out into the Universe from Oldham at the moment of my birth – perhaps in celebration, who knows? After one second, the light would have travelled a distance of 1 light second. We would mark the point in spacetime that the beam of light reached as an event at position 1 second x c on the time axis, and 1 second x c on the space axis. After 2 seconds the light beam would have travelled 2 light seconds, and so on. This lightcone, therefore, is the

Below: The spacetime diagram of my life from Albert perspective, sitting in Oldham Royal Infirmary.

Right: Me in my paddling pool on that hot, sunny day in August 1972, and filming in Ethiopia on 2 September 2009.

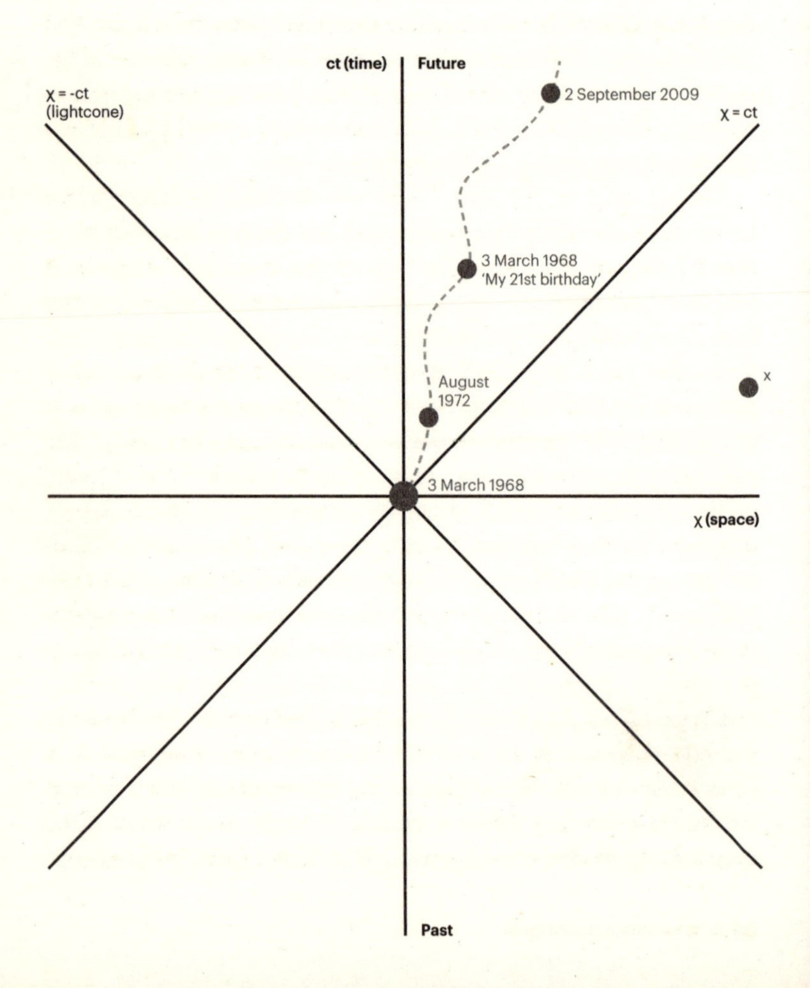

worldline of a light beam that originates at the origin of the diagram – the event of my birth. It extends all the way across the spacetime diagram at an angle of 45 degrees. My worldline wanders around inside the lightcone, as it must because nothing can travel faster than the speed of light. To see this, look at the event marked 'X' on the diagram. It is something that happened far away. Let's imagine that the event is a little alien boy on a planet 50 light years away, paddling in his swimming pool, and that, according to the Oldham watches, this event occurred at the same time as my pool adventure in 1972. We say that these events are simultaneous in the Oldham frame of reference. You'd have to travel much faster than light to get there if you were present at my birth – 50 light years in 4½ years, in fact, and this is not allowed because nothing can travel faster than light. You may have heard this many times, and wondered why. 'It's impossible to travel faster than the speed of light, and certainly not desirable as one's hat keeps blowing off', said Woody Allen. We'll gain insight into why it's not allowed in a moment.

The lightcone in the top half of the diagram is known as the future lightcone of my birth, because it marks out the region in spacetime that I could possibly visit or influence. I could not influence events outside of the lightcone in any way because I would have to travel faster than light to reach them.

There is also a lightcone in the bottom half of the diagram, which represents the time before my birth. This is called the past lightcone of my birth. My parents' worldlines must be contained within the past lightcone, because they obviously influenced it. What's more, every one of my ancestors' worldlines, stretching back to the origin of life on Earth 4 billion years ago, must also be contained within the past lightcone. No events outside the past lightcone could have influenced my birth, because no signal could have made it from them to Oldham on 3 March 1968 without travelling faster than the speed of light.

Let's now ask a question. What does this diagram look like from the point of view of our intrepid French aviator, flying past at a constant speed? We already know that she would see the Oldham watches run slow and the Oldham rulers shrink, so she would place events on my worldline at different places on her spacetime diagram.

For simplicity, let's imagine that the aviator agrees to synchronise her watches with the Oldham watches at the moment of my birth, and agrees that my birth in Oldham Royal Infirmary also occurs at position zero on her space axis. In other words, the origins of the two diagrams coincide at $t = 0$.

Although the observer sitting diligently at Oldham Royal Infirmary will not agree with the aviator on the time difference and spatial distance between events on my worldline, they will both agree on the distances in spacetime between the events, given by $\Delta s^2 = c^2\Delta t^2 - \Delta x^2$. This means that Δx and Δt must change in a very specific way, such that Δs always remains the same. The aviator's spacetime diagram is shown in the illustration opposite. Notice that the lightcones do not change, in accord with Einstein's second postulate – both the aviator and the Oldham observer must agree on the speed of light. Now look at the position of the event that represents my twenty-first birthday. We know that the aviator's clocks will tick at a different rate to the Oldham clock, and that the aviator's rulers will be a different length to the Oldham rulers. But we also know that whatever distance and time difference she measures between the '3 March 1968' event and my 'twenty-first birthday' event, they must obey the rule that Δs^2 between the events remains the same. We've drawn all the possible positions of my twenty-first birthday on the aviator's spacetime diagram as a curve. The actual position she marks will depend on how fast she flies by and in what direction. Here, we've assumed that she flies close to the speed of light in the direction of Oldham's positive x direction. Something interesting is immediately obvious. My twenty-first birthday always stays in the future lightcone of my birth. This must be the case, because my birth caused my twenty-first birthday! We'd be in trouble if, from someone else's point of view, my birthday drifted out of the lightcone of my birth and couldn't have influenced it!

So far so good. Look now, however, at the event marked 'X' – the little alien boy in his paddling pool – that lies outside the lightcone of my birth. This event must also maintain its distance in spacetime from 3 March 1968, but to do that it has to move on a different curve. Crucially, it doesn't have to stay in my future. For certain relative velocities between the aviator and Oldham, the event appears, from

Below: The spacetime diagram from the perspective of the aviator, flying at high velocity in the +x direction relative to me. I've labelled the axis as ct' and x' to emphasise that the time and space co-ordinates of the events are different in the aviator's frame of reference. The event at the origin – my birth – labelled 3 March 1968, remains at the origin because we agreed that both frames of reference have their origins coincident at t = t' = 0.

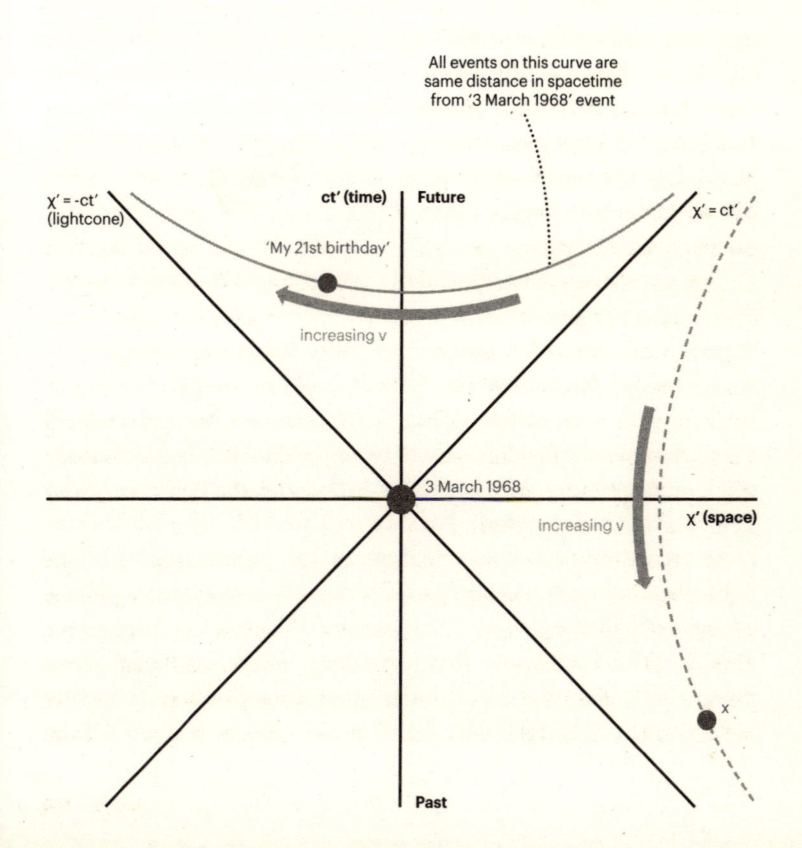

her perspective, to be in my past! This deserves an exclamation mark. The time-ordering of my birth and event X have been reversed from the perspective of the aviator. Is Einstein's beautiful theory producing nonsense? Can it really be true that the time-ordering of events in spacetime is not agreed upon by all observers? Yes it is true, but this isn't a problem, because event X always stays outside of my future and past lightcones. This means that my birth could not have influenced it, and it could not have influenced my birth. The two events are causally disconnected. This means that it doesn't actually matter what time-ordering we ascribe to such events (which are called 'spacelike separated' events) because they cannot, even in principle, have anything to do with each other. Let's give a specific example to make this clearer.

Imagine that, at the exact moment of my birth in my frame of reference, a huge explosion occurred on the Sun. The Sun is eight light minutes away, which means that the explosion cannot influence anything on Earth for at least eight minutes, which is the time it takes a light beam to travel from the Sun to the Earth. These events are 'spacelike separated', so therefore an astronaut flying past us at high speed might see the explosion happen before, or after, my birth. The time-ordering would be changed. But who cares? What difference does it make? None at all, because the events cannot influence each other.

Notice, however, that after eight minutes the shockwave from the explosion could hit the Earth and destroy Oldham, which would, to use the local vernacular, piss on my chips. Remember, though, that we are talking about events in spacetime. My birth is an event, and the explosion is an event, and my birth is outside the lightcone of the explosion and therefore cannot be stopped by it. My unfortunate death eight minutes later is another event, and that event *is* in the lightcone of the explosion. Nobody will see the time-ordering of these events reversed. Events that are in each other's past or future lightcones are known as 'timelike separated' events, and their ordering cannot be changed.

It is quite remarkable that everything works out, albeit in a rather subtle way. But there is a sting in the tail. Think about my birth event – '3 March 1968' – and event X again. In the Oldham frame of reference, event X lies in my future. In another frame of reference, event X happens simultaneously with my birth, and in the aviator's frame of reference it lies in my past. Events that happen simultaneously in one frame of reference are not simultaneous in another frame of reference. Whilst this doesn't cause problems, as we've seen, it does raise an interesting question. If there is no clear distinction between the future and the past, and indeed if an event lies in someone's future according to one observer and in their past according to another, then what do the concepts of future and past actually mean? When I was born, had event X happened or not? According to me, it hadn't. According to the aviator, it had. This suggests that, in the theory of relativity, events have an existence in spacetime beyond our local concept of past, present and future.

Let's make this more vivid. Recall that event 'X' represents a little alien boy playing in a paddling pool on a planet 50 light years away from Earth. In the Oldham frame of reference, this event happened simultaneously with my summer's day in 1972. Now look at the illustration on page 131, which shows how this event appears to the aviator travelling at high speed relative to me. There exist frames of reference in which the alien boy's paddling pool day is in my past, and my entire life, including my paddling pool day, is in his future. My summer's day hasn't happened yet. It's out there in spacetime, in his future, albeit in a region of spacetime inaccessible to him. From my perspective, my 1972 paddling pool day is in my memory. I remember it with fondness. Surely it's gone, hasn't it?

If we take Einstein's theory of relativity at face value, there is no sense in which the past has happened and the future is yet to happen. A spacelike separated event can be in someone's future from one perspective, and in their past from another. This doesn't matter in the sense that such events can have no influence on each other, provided that nothing can travel faster than the speed of light. This is why the speed of light as a universal speed limit is so important in relativity. It protects cause and effect. But this behaviour does raise the question of whether all events that can happen and have happened in the history of the Universe are, in some sense, 'out there'. This idea is known as the Block Universe. Spacetime can be pictured as a four-dimensional blob over which we move, encountering the events on our worldline as we go. We are forced to move over the blob at the speed of light, which from our own personal perspective means that we have to move into the future at a speed of 1 second per second.[2] You have to get old because of the geometry of spacetime.

We should emphasise that, while the Block Universe is a consequence of relativity, it is not necessarily correct. We know that relativity is not fully consistent with quantum theory, and most physicists hope and expect that a quantum theory of spacetime will be developed at some point. Whether this will allow for a more intuitive picture of past and future is unknown. We must always remember that physical theories such as relativity are models of reality that produce predictions that agree with experiment – a test which both the Special and General theories have passed with flying colours for over a century. Is the Block Universe actually real, or just an artefact of Einstein's model? Who knows? But I think its implications are at the very least worth thinking about. On the downside, there is no free will in the Block Universe. All the events in our future 'exist', waiting for us to barrel along our worldline to intersect them. I don't care personally whether I have free will or not. It makes no difference to me. But I find the other side of the coin quite wonderful. In the Block Universe, the past is also out there. My idyllic summer's day in 1972, with my Mum and Dad and sister, doesn't exist only in my memory. It hasn't gone, although I can never revisit it. It is still there; all those people, all those moments, always and forever, somewhere in spacetime. I love that.

2 You can see this from the definition of distance in spacetime. Set $\Delta x = 0$, because you are in your own rest frame, and note that $\Delta s / \Delta t = c$.

Spacetime calculations

Monet and the aviator

We can use Einstein's two postulates to show why it is that the aviator and Monet measure different intervals of time between any pair of events. This is surely one of the most bizarre ideas ever to come out of a human being's head. It is all the more bizarre for being demonstrably correct. The argument is surprisingly simple. First let us imagine a special type of clock – at the end we will show that the argument must work for any type of clock, but for now we will consider a 'light clock'. A light clock is made up of two parallel mirrors with a beam of light bouncing back and forth between them. Suppose that the two mirrors are a distance d apart. If light travels at a speed c it will take a time $t=2d/c$ for the light to travel from one mirror to the other and back again, as determined by someone who is holding the clock (more formally, we might say 'by someone who is at rest relative to the clock'). Let us refer to the person holding this clock as (and here we will not bother exercising our imagination) 'person A'. Now let's introduce a second person: 'person B'. If person A and person B are both at rest relative to each other then both will clearly agree on how long the light clock takes to tick (let's call one tick of the clock the time it takes for the light to make one round-trip, i.e. $t=2d/c$). Pre-Einstein, and according to common sense, we'd say that the clock takes this time t to tick, regardless of what it is doing or who is doing the measuring. But that is wrong, as we are about to show.

To see how time is not absolute, let's put person A and their clock on a train (Einstein often used trains to explain his theories), and person B on the platform. Now let us consider how the clock is understood by person B. The top illustration shows the path taken by the light as it makes one tick of the clock.

According to person B, the clock moves a distance equal to vt' in one tick, where v is the speed of the train and t' is the duration of the tick. At this stage we will resist the temptation to say that t' (the time of one tick of the light clock according to person B) is equal to t (the time of one tick according to person A). From the figure we can see the path that the light beam traces out as it moves up and down. Obviously, the light travels further according to person B than it does according to person A. Using Pythagoras's Theorem, the distance the light travels according to person B is $2\sqrt{(vt'/2)^2+d^2}$, whilst for person A it is just $2d$ (notice that it would be just $2d$ for person B if $v=0$, i.e. if the train isn't moving). The fact that the light travels further according to person B is not by itself anything to get excited about, because the train is moving. The next step is the shocker.

Einstein's second postulate states that the speed of light in a vacuum is the same in all inertial frames of reference. It follows that person B must agree that the light moves at a speed c. If the light moves at speed c according to both A and B, and if the light travels further according to person B than it does according to person A, then it follows that the light must take longer to make the round trip according to person B than it does according to person A. This is worth re-reading and thinking about, because it is surprising.

We have just proven that, if Einstein's second postulate is correct, it logically follows that the light clock ticks more slowly according to person B (who is on the platform) than it does according to person A (who is on the train). Since we went to the trouble of invoking Pythagoras and a little algebra to write down how far the light travels in one tick according to person B, we can easily write down by how much the moving clock slows down according to person B. The time taken for one tick, according to person B, is the distance the light travels in one tick, divided by the speed of light c, i.e. $t'=2\sqrt{(vt'/2)^2+d^2}/c$. Notice that the time we want to know (t') is on both sides of this equation, which means we have to re-arrange the equation using some low-level algebra. Squaring both sides of the equation gives $t'^2=4(vt'/2)^2/c^2+4d^2/c^2$, which can be rearranged to read $t'^2(1-v^2/c^2)=4d^2/c^2$. Now we can write down what t' is in terms of v, d and c. It is just $t'=2d/c/\sqrt{1-v^2/c^2}$. And since $t=2d/c$ we can write

Below: Monet and the aviator. **Bottom:** Hyperbola.

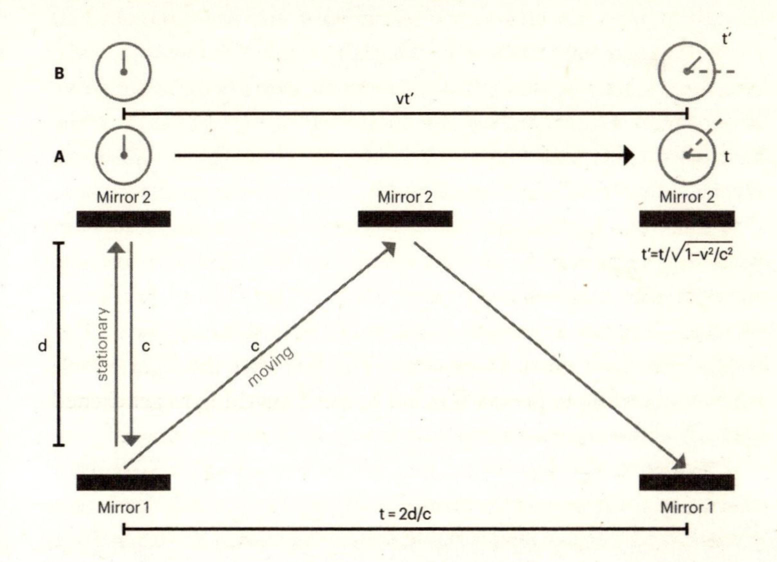

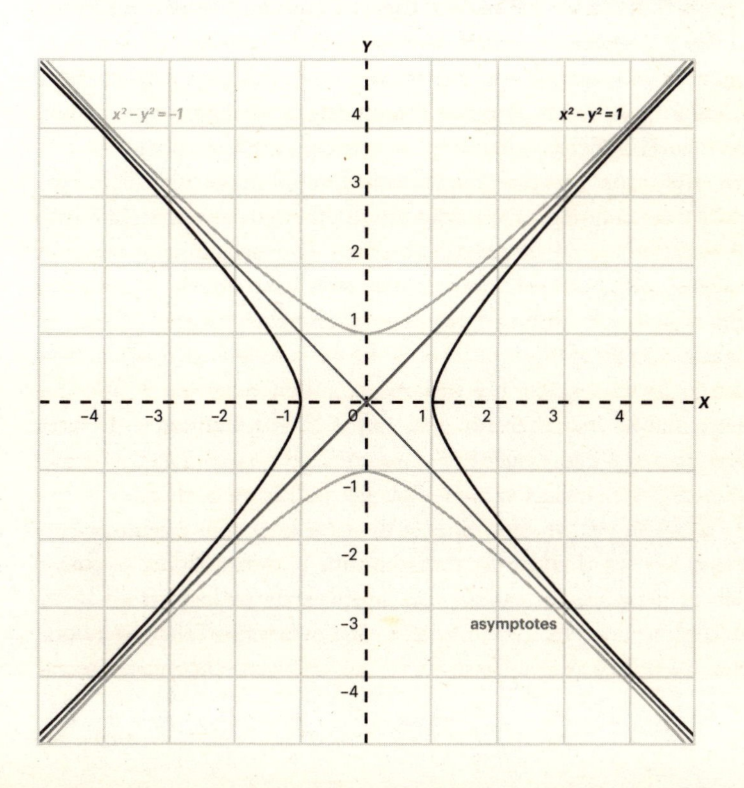

down that $t' = t/\sqrt{1 - v^2/c^2}$. And that is our final answer. So long as v is smaller than c, the square root makes sense and t' is always bigger than t, which means that the person on the platform must conclude that the person on the train is holding a clock which is taking longer to tick than it would if the clock were not moving. As an aside, the factor $1/\sqrt{1 - v^2/c^2}$ appears very often in relativity, and is known as the Lorentz factor or Gamma factor, and given the symbol γ.

Before we start to claim that the world is an amazing place, we ought to convince ourselves that the result we just found is not just some peculiar feature of light clocks. First, let's be clear about why we chose to think about a light clock in the first place. We did that because we could make direct use of Einstein's postulate about the speed of light being the same in all inertial frames. If we had been thinking about pendulum clocks then we couldn't have exploited that postulate so easily. But, with more work, we could have done this calculation using pendulum clocks, or heart beats, or any other type of clock, and the conclusion would have been exactly the same. You can see that this has to be true if Einstein's second postulate is correct:

The laws of physics are the same in all inertial frames of reference,

or, more colloquially, 'it is impossible to tell who is moving and who is standing still'. Suppose that the slowing of the light clock is some peculiarity of light clocks and that it doesn't apply to other clocks. If that were true, person A (on the train) would notice that their light clock was running slow compared to their wristwatch. But that observation would be enough for them to conclude that they are moving, which would be in conflict with Einstein's first postulate. The only way to keep that postulate alive is to say that if person A's light clock takes longer to tick according to person B then so too must person A's wristwatch. It is time to acknowledge that the world is far more remarkable than we had any right to suppose. We have demonstrated that, if Einstein's two postulates are correct, two people in motion with respect to each other age at different rates.

To finish off, we can compute the time interval between the two events we discussed in the text: the time between Monet placing a dab of paint on his canvas and the turn of the lock in his door. According to Monet's timepiece, the interval was eight hours. But, according to the formula we just derived, the time interval

Below: Scott Carpenter signed this photograph for my son, George, shortly before Carpenter died, aged 88, in October 2013.

Right: Wilson Bentley absorbed in capturing unique and delicate images of snowflakes on film in Vermont in 1885.

Below: These captivating images, taken by Wilson Bentley through a light microscope attached to his camera, reveal the uniqueness of each snowflake.

Below: Every winter the warm waters of Florida are home to one of Nature's apparently less elegant shapes. The caveat is important, because the clumsy-looking manatee is as well adapted to its environment as the most aesthetically refined butterfly.

Coca-Cola

ELS PETITS
ENS FAN
Damm

Previous page : The teams in action building the *Castells*, human towers that defy gravity – and fear!

Below: One of Claude Monet's most famous works, *Coquelicots* (Poppies), captures a perfect moment in time, on one summer's day in the French countryside in 1873.

Right: This grainy black-and-white image occupies an historic place in the archives of meteorology. Taken by NASA's TIROS-3 satellite, it is the first time a hurricane was discovered using satellite imagery and is one of the first photographs of a tropical cyclone from space.

Below: The Voyager probes were launched in 1977, destined never to return from their mission to explore the outer reaches of the Solar System. In their travels around interstellar space they have sent back images of Jupiter and Saturn.

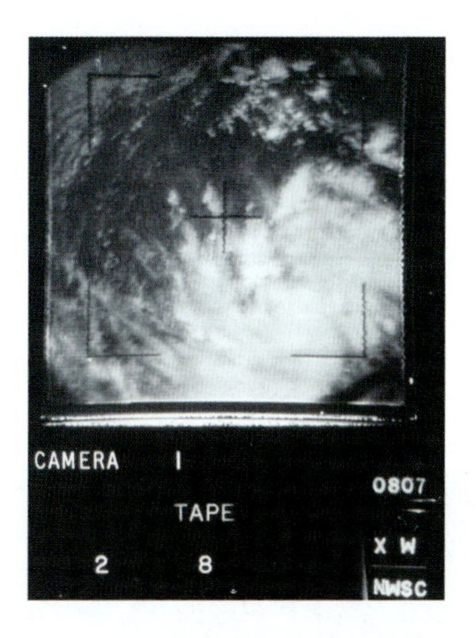

Below: The effect of the variance of gravitational pull and the Centrifugal Force of the Moon can be witnessed all over the globe in our tides. These tides change the appearance of our coastal landscapes every day, and we now have the technology to monitor them and predict them to be able to use that information for maritime purposes – as well as leisure pursuits such as surfing!

Left: At the heart of Mount Ijen volcano lies the largest acidic crater in the world, serene and beautiful, but one of the most extreme natural environments on Earth.

Below: The discovery of zircon crystals in the Jack Hills gives us evidence that the atmosphere 4.4. billion years ago was very similar to that of today.

Right: Kamal al-Din al-Farisi's beautiful manuscript explaining the mathematical explanation of the formation of a rainbow.

Below: The ice fountains of Enceladus, photographed by Cassini on her final flyby of the Moon in October 2015.

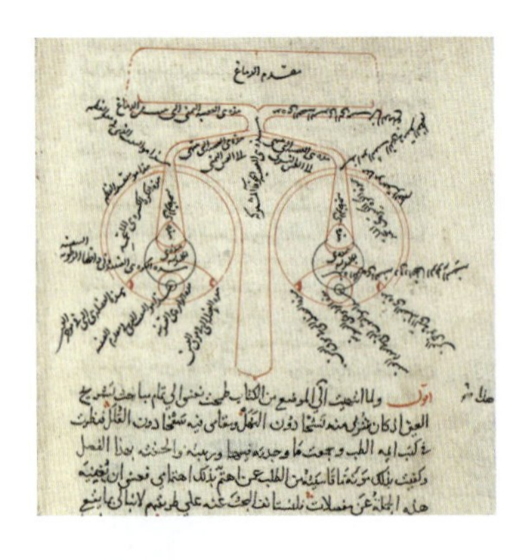

Left: Prism demonstrating refraction and reflection effects. A beam of white light strikes the prism and is dispersed onto the opposite face. Some of the light is refracted again, exiting the prism and forming the spectrum.

Below: Coronal mass ejections and solar flares demonstrate the sheer energy of the Sun – one such explosion has the power of one billion hydrogen bombs.

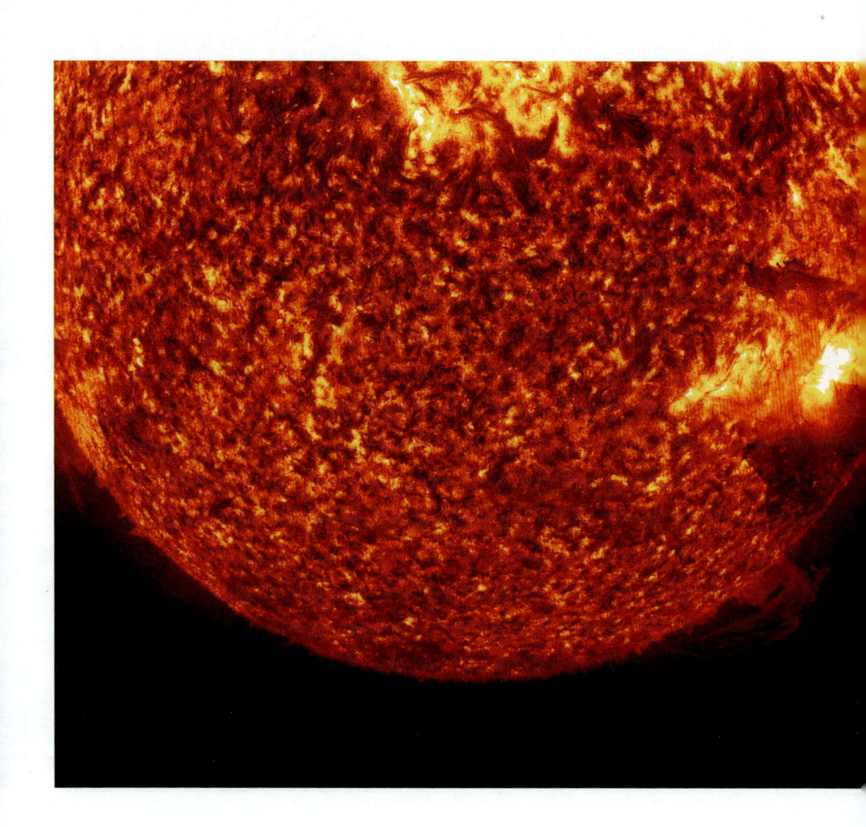

Left: The space-age Super-Kamiokande Neutrino Detector is housed under Mount Ikeno in Japan, and here, 1000 metres underground, scientists study solar and atmospheric neutrinos.

Below: The white dwarf Sirius B shines bright while the Sun gradually radiates its heat away until it is left as a darkening ember, a black dwarf.

Right: The 'White Marble' – an Arctic view of Earth from the Blue Marble 2012, taken by NASA's Suomi-NPP Satellite.

Below: The thin blue line that separates our planet from the vacuum of space, as seen from the ISS.

between the same two events as measured by the aviator is 8 hours/ $\sqrt{1-1/4}$=9.24 hours.

Hyperbola

The distance between two events in spacetime, $\Delta s^2=c^2\Delta t^2-\Delta x^2$. Physically, for any two events, although the distance in space Δx and the distance in time Δt will change when measured by observers in different frames of reference, they must change such that the distance in spacetime, Δs, remains constant. Mathematically, this is the equation of a curve known as a hyperbola. Let's consider a specific example, equivalent to setting the distance between two events in spacetime to be 1 unit. There are two versions of this 'unit hyperbola', which have the equations $x^2-y^2=1$ and $x^2-y^2=-1$ These are shown in the bottom illustration on page 137. If x^2-y^2 is equal to 1, there are two curves; one in the upper half plane and one in the lower half plane. This is the situation for events that are 'timelike separated', i.e. causally connected to each other. For such events, the distance in space between the events, Δx^2, is always less than the distance light could have travelled during the time interval between the events $c^2\Delta t^2$. The spacetime interval Δs^2 will therefore always be positive. If the events are not causally connected, which is to say that the spatial distance between them Δx^2 is always greater than the distance light could have travelled during the time interval between them $c^2\Delta t^2$, then we have the curve $x^2-y^2=-1$. These events are known as 'spacelike separated'. As we note in the text, the important point is that when we change between inertial frames, the events slide around the spacetime diagram on these four curves, but never hop between them: causally connected events (timelike separated) *always* have their time-ordering persevered, but causally disconnected events (spacelike separated) need not.

ELEM

ENTS

The moth and the flame

How did life begin? I think this is one of the two most interesting questions in science, and the most important question in the history of human thought. Cathedrals have been built, wars have been fought and empires have risen and fallen as innumerable demagogues have sought universal agreement for their guesses.

How did the Universe begin? I think this is the other interesting scientific question, but we know less about it at the moment. There are speculative theories that suggest the Universe could be eternal, and that there was no beginning. If this is the case, then the Universe has always existed and the question is answered: It didn't begin. Whether or not this would be a satisfying answer is up to you. I'd be comfortable with it.

Irrespective of what happened at the beginning of time, we know there was a period 13.8 billion years ago when the part of the Universe that we can see today, containing over 350 billion large galaxies and with a diameter of over 90 billion light years, was compressed into a region of space smaller than a single atom. No life could have existed in such extreme physical conditions, so even if the Universe existed in some form before the Big Bang, it's safe to say that no complex physical structures would have made it through. The observable Universe must therefore have been devoid of life at some point in the past, and life must have begun spontaneously somewhere within it, at some point during the last 13.8 billion years. The word 'spontaneously' is worth defining here, because it crops up a lot in discussions about the origin of life. In saying that life appeared spontaneously, we are asserting that life is a physical process that emerged as a result of the action of the laws of Nature. If we say that the Earth formed spontaneously, we mean that nobody built it; by saying that living things appeared spontaneously, we mean the same thing.

The first atomic nuclei formed in the initial minute or so following the Big Bang, and the first atoms formed in large numbers when the Universe was 380,000 years old. The first stars ignited around 100 million years later, and these assembled the first carbon atoms. It is unlikely that the rich chemistry of life could have begun spontaneously without carbon, oxygen and a handful of the heavier elements beyond the hydrogen and helium atoms that existed before the stars. Life could have got going anywhere in the Universe after this time, and may well have done; we don't know.

The Earth formed 4.54 billion years ago out of the primordial cloud around the young Sun. It is safe to assume that there was no life on Earth in the early years following her formation; the conditions were too violent and changeable. There is good evidence that life had gained a foothold on Earth around 3.5 billion years ago, and possibly much earlier – we'll discuss this evidence later. Therefore, we will assume that Earth was once a lifeless world, and that living things appeared at some point in the first billion years after its formation.

Nothing that we've said in these opening paragraphs is controversial from a scientific perspective, but there is one assumption we'll make which is at least contestable. We will assume that life began on Earth, rather than arrived here from space. Since we have discovered no life on planets beyond Earth, this is reasonable, but it is possible that life began on Mars, or perhaps even on comets in the outer Solar System, and was delivered to Earth by impacts from space. This theory is known as panspermia. Unlikely as it may sound, it is a testable theory, and that puts it firmly in the realm of science.

It is certainly possible that we could discover life on Mars over the next few decades, and if the microbial Martians share our biochemistry and our genetic code, we might be forced to postulate a common origin on either planet, or perhaps somewhere else in the Solar System. This would make the search for the origin of life more difficult, because it is far easier to explore our own planet's deep history than it is to explore the history of another world. The only way to find out is to do the science, and this is one of the reasons why we send spacecraft to Mars and the potentially life-supporting moons of Jupiter and Saturn. It goes without saying, however, that we shouldn't stop searching for the origin of life on Earth whilst we build spacecraft to search for life beyond it.

Under the assumption that life began on Earth, it must have been the case that the basic chemistry of life existed on our planet before living things emerged, and that sometime and somewhere chemistry became biology. There is no precise definition of what 'becoming biology' means, but it is worth emphasising that biology is just a word for (very) complex chemistry. Living things are constructed from the same set of chemical elements as inanimate things, and they obey the same laws of Nature. In this sense we can assert that the Earth is our ancestor and creator, and we would like to know how, where and when the transition from geochemistry to biochemistry occurred.

Chemistry is all about the movement of electrons

Living things are made out of simple building blocks with complex interactions. This is obvious in one sense, because everything in the Universe is made out of simple building blocks with complex interactions. We saw in Chapter One that, at the deep subatomic level, there are only three building blocks of everyday matter: up quarks, down quarks and electrons. At a higher level, there are protons and neutrons, and higher still are the 92 naturally occurring chemical elements found on Earth. Nobody other than chemistry students or Tom Lehrer remembers all their names, but I am sure virtually every reader of this book will know that they can be laid out in a pattern known as the periodic table. Each element has a different number of protons in its atomic nucleus, and an equal number of electrons surrounding it; hydrogen has 1 proton and 1 electron, carbon has 6 protons and 6 electrons, and so on. Chemical and biochemical processes are about the sharing and transfer of electrons between elements, allowing molecules to be formed and broken apart.

All the elements beyond element 92, Uranium, were constructed in laboratories, usually by the bombardment of heavy atomic nuclei by neutrons, or by forcing lighter elements to fuse together. The heaviest goes by the name of Ununoctium, and has 118 protons and 176 neutrons in its nucleus. This exotic nucleus is highly unstable and lives for less than a millisecond.

The periodic table is more than just a pictorial arrangement of the elements; it is the key to understanding how and why elements react together to form molecules. The vertical columns, called groups, contain elements with similar chemical properties. The reason for this is that all the elements in a particular group have the same number of electrons in their outer shells, and it is these electrons that are available for sharing or donating to other atoms; they are the particles that make chemistry happen. We explored the structure of oxygen and hydrogen atoms in some detail in Chapter One. Hydrogen has a single electron. Oxygen has 8 electrons, of which 2 sit close into the nucleus and play no part in chemical reactions. The remaining 6 populate its outer shell, and 2 of these are on their own. Oxygen would dearly like 2 more to complete its outer shell, and it will take them from other atoms given half a chance. This is why hydrogen and oxygen will get together, given a very tiny nudge, to form H_2O. I am aware that this anthropomorphic language is a bit unscientific, but to be honest I don't care. I hope it makes the point, which is that it is the arrangement of electrons inside the atoms of the different chemical elements that leads to chemistry.

Sulphur has 16 electrons, of which 10 fit close into the nucleus, leaving 6 in its outer shell, just like oxygen. This means that sulphur, just like oxygen, will grab 2 electrons from other atoms if it can. In the presence of hydrogen it will form the molecule hydrogen sulphide, H_2S, one of the constituents of the Earth's primordial atmosphere. Carbon has 4 electrons in its outer shell, and it shares them all to form compounds like carbon dioxide, CO_2. Below carbon in the periodic table is silicon, which also has 4 electrons in its outer shell, and it forms similar compounds such as silicon dioxide, SiO_2, and so on. As we discussed in Chapter One, electrons are arranged in this highly structured way around atomic nuclei in accord with quantum theory, which is part of the fundamental set of the laws of Nature. The important point is that electrons can be transferred or shared between the atoms of different elements, and this is what drives the formation of molecules. Chemistry is all about the movement of electrons, and the movement of electrons can lead to complexity.

Frankenstein's monsters

Any discussions about the historical details of the emergence of the modern scientific worldview are at least partly subjective, and academics spend their careers exploring the subject. If a working physicist is asked to identify the first recognisably modern scientific theory, it is likely they will point to Newton's *Principia* of 1687, not least because Newtonian physics is still in use today and is taught as part of a twenty-first-century degree course. If you want to send a spacecraft to Pluto, you make the navigational calculations using Newton's laws. *Principia* provides a complete and self-consistent model for the geometry and dynamics of the Solar System, with the Sun at the centre and the Earth orbiting around with the rest of the planets. If there was anybody left in 1687 arguing that the Earth occupies a unique physical place, stationary beneath the stars at the centre of creation, they would certainly have had to shut up when presented with a copy of the *Principia*.

The transition from furious debate about an Earth-centred cosmos to the near-universal acceptance of our physical demotion was relatively rapid once Copernicus and others had opened the intellectual floodgates during the sixteenth century. It's easy to overlook the philosophical, intellectual and theological storms that the merger between observational astronomy and theoretical physics precipitated.

Newton was born in 1643, only a year after Galileo died, and Galileo famously encountered quite serious resistance to his support for an orbiting Earth. I've written of my admiration for Kepler's beautiful writing in *On the Six-cornered Snowflake*; a distinctly modern voice suffused with wit, curiosity and a careful approach to the exploration of Nature echoes down the centuries. Galileo was possessed of a similar confidence and amusing turn of phrase. Here he is, writing to Kepler in 1610, taking a magnificently belligerent swipe at the Earth-centred-Universe lobby;

'My dear Kepler, I wish that we might laugh at the remarkable stupidity of the common herd. What do you have to say about the principal philosophers of this academy who are filled with the stubbornness of an asp and do not want to look at either the planets, the moon or the telescope, even though I have freely and deliberately offered them the opportunity a thousand times? Truly, just as the asp stops its ears, so do these philosophers shut their eyes to the light of truth.'

Galileo would have been at home on Twitter.

Our physical demotion from the centre of all things was well established by the end of the seventeenth century and has continued relentlessly ever since. The entirety of our observable universe is an irrelevant pocket of dust in the wider cosmos, which extends way beyond the visible horizon and is conceivably infinite in extent, and I think society has come to terms with this sort of physical irrelevance. It's hard to look at the Hubble Ultra Deep Field Image, containing over ten thousand galaxies in a piece of the night sky you'd cover very comfortably with an outstretched thumb, and feel important. Our spiritual demotion, however, is an entirely different matter. By spiritual demotion, I mean the realisation that our very existence has no more significance than our physical location. This is surely the case if life is the inevitable result of the action of the same set of natural laws that formed the stars and planets. Earth must be one of countless billions of living worlds in the Milky Way galaxy alone. This is absolutely not to suggest that our civilisation is not worth celebrating and fighting to preserve – it is my view that civilisations may be extremely rare, even if life is common.

It is possible to make an argument that there are only a handful of civilisations in the Milky Way galaxy today – perhaps we are the only one? – and this makes planet Earth a rare and valuable natural phenomenon. Value is in the eye of the beholder, and whilst it would be irrational to attach any universal significance to our temporary existence in a possibly infinite cosmos, I do not see any contradictions raised by the use of the word. Intelligence brings meaning to the

Universe, albeit locally and temporarily. Our existence obviously means something to you and me, and I do not accept that our physical irrelevance and temporal transience devalues our lives one iota.

This is territory over which philosophy and theology still claim partial dominion, but science inevitably wanders into this intellectual no-man's-land because to discuss the origin of life is to discuss the origin of humanity, with all the intellectual baggage that brings. Is it possible that my feelings, my morality, my hopes, fears and loves will be explained by some future biological *Principia*, as surely as the motions of the planets are explained by Newton? Is my apparent freedom of will an illusion resulting from the action of deterministic physical laws? Have those shadows on the wall of Plato's cave deceived me into thinking I am special, as the rotating stars on the celestial sphere once conspired to? These are questions of a different emotional magnitude from those about our physical location in the Universe, and it is absolutely clear that they remain unsettled in the minds of many, although this is irrelevant in the sense that the veracity of a scientific theory is not decided by referendum.

I think our modern understanding of biology and the scientific search for the origin of life must form an essential part of any serious philosophical debate about the meaning and value of human life. Having said that, given the overwhelming visceral force of our individual experience of living, it is perhaps not so paradoxical that the place of an individual human being in the Universe is still vigorously debated, whilst the physical position and significance of our planet is not.

The scientific quest to explain the origin of life became fashionable in spectacular style at the turn of the nineteenth century. Advances in surgery, pioneered by anatomists and surgeons such as John Hunter, dovetailed with the discoveries in electricity and magnetism pioneered by Faraday and his contemporaries, and coalesced into the search for a 'vital principle' – the animating force that separates living from inanimate matter. As far back as 1780, Luigi Galvani had been causing dead frogs' legs to twitch 'back into life' by passing electrical currents through them, an approach that reached an infamous zenith in the hands of Giovanni Aldini, Galvani's nephew, on 17 January 1803. Aldini procured the corpse of convicted murderer George

Below: These plates show Giovanni Aldini and colleagues in action, experimenting on both human cadavers and those of animals in an attempt to reanimate the corpses.

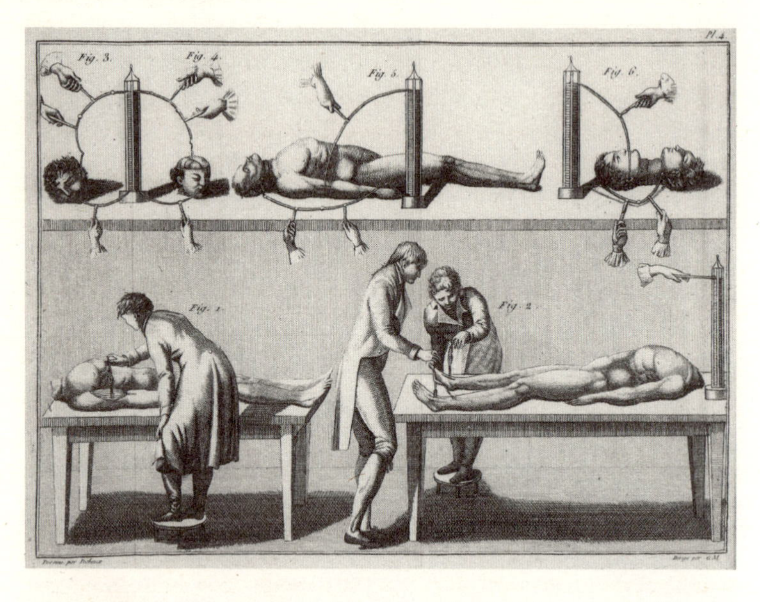

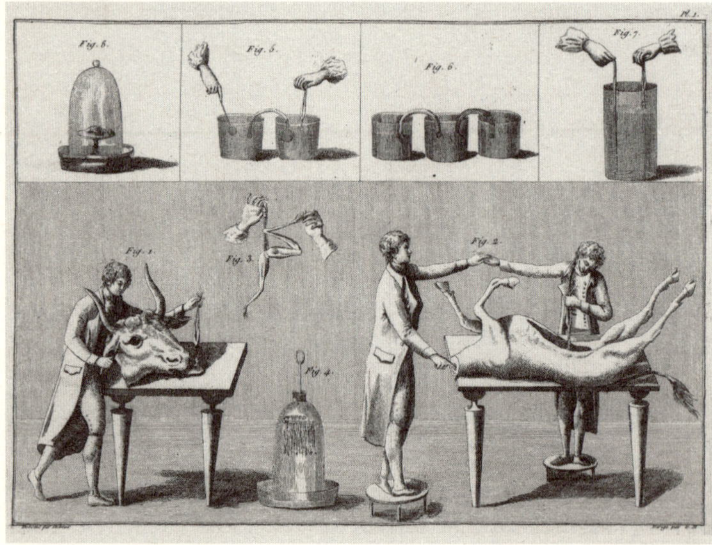

Forster, fresh from the gallows at Newgate Prison, and attempted to reanimate it live on stage in front of an astonished audience. The *Newgate Calendar* reported it thus:

'On the first application of the arcs to the face, the jaws of the deceased criminal began to quiver, and the adjoining muscles were horribly contorted, and the left eye was actually opened. The arms alternately rose and fell, the fists clenched and beat violently the table on which the body lay, natural respiration was artificially established…A lighted candle placed before the mouth was several times extinguished… Vitality might have been fully restored if many ulterior circumstances had not rendered this inappropriate.'

The growing fascination and disquiet surrounding the scientific push into such fraught territory was captured most famously in Mary Shelley's novel *Frankenstein; or, The Modern Prometheus.* A writer knows they have done their job when reviews are polarised; it means they are operating in contested intellectual terrain. The poet and novelist Sir Walter Scott reviewed the novel favourably, although contemporary gossip had it that he may have been the author – Shelley had published the work anonymously. A critic writing for the *Quarterly Review* described it as 'a tissue of horrible and disgusting absurdity'. I must remember that and aspire to write something worthy of such a label.

Today, Frankenstein's 'monster' is a creature of B-movie horror films, but in Shelley's novel the animated being is an articulate and moving voice.

'And what was I? Of my creation and my creator I was absolutely ignorant? … No father had watched my infant days, no mother had blessed me with smiles and caresses …'

Two centuries later, with the whole of modern medical science, evolutionary biology and genetics to draw upon, we are still unable to reach an accommodation between our desire to discover a reason for our creation and the scientific consensus that no such reason exists beyond the inevitable action of the laws of Nature on a young, active planet. The most interesting questions are those that demand a resolution between apparently irreconcilable positions.

On the Origin of Species

A framework to make sense of life on Earth

Forty years after the publication of *Frankenstein*, in November 1859, Charles Darwin's *On the Origin of Species* provided the necessary conceptual framework for the scientific exploration of the origin of life, much as Einstein's Theory of General Relativity provided the necessary conceptual framework for the study of the origin of the Universe. Darwin recognised that the great diversity of different species on Earth, the endless forms most beautiful, as he memorably called them, are related to one another. We now know this is correct, but for Darwin it was a radical proposal, indeed an act of genius, given the evidence available at the time. He was able to reach this conclusion by proposing a mechanism for new species to emerge from older ones: Evolution by natural selection.

There will be genetic variation in a population, which we now know to be caused by random mutations in the genetic code, the shuffling of genes by sex and a host of other mechanisms. Because organisms pass on genes to their offspring, combinations of genes that make an organism more likely to survive long enough to reproduce will become more common in a population. In this way, populations are shaped very rapidly by their interactions with the environment and with other living things. If populations become separated and have little or no interaction with each other, these processes drive them apart genetically, physically and behaviourally, and this is how new species emerge. Separation can be geographical, as in the case of the unique flora and fauna found on islands such as Madagascar, or it can result from different environmental niches opening up in a given location.

Once it is accepted that species do not appear fully formed, do not remain unchanged, and will inevitably evolve into new species if they are separated in time and space and exposed to different selection pressures, it is at least a possibility that all living things might have shared a common ancestor at some point in the past. As Darwin wrote: 'Therefore I should infer from analogy that probably all organic beings which have ever lived on this earth have descended from some one primordial form, into which life was first breathed.'

Darwin didn't know whether this was correct, but he knew it was possible. In a letter to his friend and colleague Joseph Hooker, he went further, speculating about the origin of life on Earth in some primordial 'warm little pond'. It is said his imagination was fired after reading about an experiment demonstrating that some moulds could survive boiling.

> *'It is often said that all the conditions for the first production of a living organism are now present, which could ever have been present. But if (and oh! what a big if!) we could conceive in some warm little pond, with all sorts of ammonia and phosphoric salts, light, heat, electricity, &c., present, that a protein compound was chemically formed ready to undergo still more complex changes, at the present day such matter would be instantly devoured, or absorbed, which would not have been the case before living creatures were formed.'*

It's hard to overstate how bold and visionary Charles Darwin was. This was 1859, three years before Lord Kelvin declared that the Sun and therefore the Earth could be no more than 30 million years old, based on the known physics of the day. We will discuss the resolution to this problem in Chapter Four. It's very difficult to imagine how some form of primitive single-celled organism could emerge from inanimate building blocks and then be transformed into a human being by the action of natural selection in a few million years. A few billion, on the other hand, is an entirely different matter. Darwin, quite rightly as it turned out, chose to ignore the physicists, and as the years progressed, evidence mounted for his idea of a warm little pond, a geological incubator within which 'the first creature, the progenitor of innumerable extinct and living descendants, was created'.

The oldest life on Earth

If we are to build a scientific picture of Darwin's warm little pond in the broadest sense, as the incubator for the first life on Earth, we need to understand what the conditions on our planet were like when life began. We also need to look for evidence of the earliest life, so we know how far back in time we have to go. This is non-trivial, to use a favourite phrase of physicists, because we can be sure that these events happened a long time ago.

There is strong evidence that life existed 3.4 billion years ago from microfossils laid down in sandstone deposits at Marble Bar in Western Australia. The fossilised objects in the photographs certainly look like the remains of living cells, but visuals can be deceiving. Fortunately, it is possible to carry out a chemical analysis of these ancient structures, and signatures characteristic of a biological origin have been found. The concentrations of different isotopes of carbon in the structures can be used as a biomarker. Carbon has 6 protons in its atomic nucleus, and the most commonly occurring form also has 6 neutrons. This is known as the carbon-12 isotope. There is another naturally occurring form of carbon with 7 neutrons in the nucleus, known as carbon-13. Life prefers to use carbon-12, so therefore carbon deposits formed by biological processes are expected to show an excess of the lighter isotope. This is the case for the Marble Bar structures. There are also high concentrations of nitrogen in the proposed cell walls, again indicative of biological origin. Most biologists accept that these and other samples from different sites constitute strong evidence that single-celled organisms known as prokaryotes were abundant on Earth 3.4 billion years ago.

The oldest known objects on Earth were discovered in a remote region of Western Australia, north of the city of Perth, and remarkably they contain evidence of biology. Zircons are crystals found in igneous (volcanic) rocks. Despite being no bigger than a grain of sand and generally uninspiring to view, they are of immense scientific value because they are near-indestructible time capsules that carry their own internal clocks.

As the zircons form from cooling lava, tiny samples of atmospheric gases are sealed inside. Radioactive uranium atoms are also incorporated into the crystal structure, and using a highly accurate technique known as uranium-lead dating, the time since their formation can be measured to within a few million years. A sample from Erawandoo Hill, in the Jack Hills range, was recently dated at 4,404 +/- 8 million years old, making it the oldest object of terrestrial origin ever to be discovered. The Earth's age is measured to be 4540 +/- 50 million years old, so these crystals formed as the young Earth was cooling. Analysis of the trapped gases produced surprising results, challenging the commonly held picture of the young Earth as a Hadean hell of seething lava and toxic atmospheric gases. Earth was already a blue planet when some of the more ancient zircons formed, with liquid water on the surface. Atmospheric oxygen levels were low, which is unsurprising because photosynthesis is the primary source of atmospheric oxygen, but other than this, the primordial atmosphere appears to have been similar to that of today, with abundant nitrogen, carbon dioxide and water vapour as well as increased sulphur dioxide levels from the active volcanoes. This new evidence suggests that the very young Earth was a world of moderate temperatures, stable oceans and familiar air.

In November 2015, a team from UCLA and Stanford universities published a paper based on an analysis of over 10,000 zircons from the Jack Hills region, formed over 4.1 billion years ago.[1] The zircons contained carbon deposits, and in common with the Marble Bar fossils, the ratio of carbon-12 to carbon-13 is suggestive of a biological origin. This is a surprising result; as team member Mark Harrison noted, the idea that life existed on Earth a billion years after its formation would have been near heretical only twenty years ago. If the interpretation of the new zircon results is correct, in Harrison's words, 'life may have started almost instantaneously', and a terrestrial biosphere may have been well established 4.1 billion years ago.

'THE LINK BETWEEN LIVING AND DEAD MATTER IS SOMEWHERE BETWEEN A CELL AND AN ATOM.'

– J. B. S. HALDANE, 'THE ORIGIN OF LIFE', 1929

The mounting evidence that life began on Earth pretty much as soon as it could lends a sense of inevitability to the emergence of biology from chemistry. This is, of course, a subjective judgement, because we have only a single planet as evidence, and firm conclusions are difficult to draw from sample sizes of one. This is another reason why the searches for life on Mars, or the moons of Jupiter or Saturn, or on exoplanets around nearby stars, are so important. We'll have more to say about the study of planets beyond the Solar System in Chapter Four. That said, the observation that life may have emerged 'almost instantaneously' is an interesting one. Christian de Duve, the Belgian Nobel Prize-winning biochemist, argued that chemical reactions tend to proceed very quickly or not at all. Since biology is chemistry, then given the right conditions it follows that biology should happen very quickly or not at all, and the evidence from the zircons of Western Australia seems to point in this direction.

[1] Elizabeth A. Bell, 14518–14521, doi: 10.1073/pnas.1517557112

A warm little pond?

The idea that life might simply have 'popped into existence' from a soup of inanimate ingredients may seem either plausible or ridiculous to you. In the mid-nineteenth century, some of the great names in science were firmly on the side of ridiculous. The idea that life could arise from dead matter, known as spontaneous generation, had been discussed since the time of Aristotle. This is not unreasonable, because maggots appear to emerge fully formed from rotting meat. A series of experiments, most famously by Louis Pasteur and, later, John Tyndall, appeared to refute this notion, and led to the so-called law of biogenesis; the idea that living things can be produced only from other living things. As Pasteur wrote of his experiment, rather immodestly, in 1864, 'Never will the doctrine of spontaneous generation recover from the mortal blow struck by this simple experiment' and 'those who think otherwise have been deluded by their poorly conducted experiments, full of errors they neither knew how to perceive, nor how to avoid'.

This is to confuse the spontaneous emergence of an intact, complex organism like a maggot or even a bacterium cell with the spontaneous emergence of life's basic biochemistry – which is perhaps understandable if there is no known mechanism for complex living things to emerge from simpler forms. It is one of the many treasures of Darwin's theory of evolution by natural selection that it describes such a mechanism, and as Darwin himself realised, this makes the spontaneous emergence of life at least a possibility; life can start simple. I don't know why Pasteur didn't notice that *On the Origin of Species*, which was published five years before he made his definitive statement, provides a way out. I leave that judgement to the historians; perhaps he hadn't read it.

Pasteur's powerful dismissal of spontaneous generation may have had an effect, because the search for the origin of life on Earth seemed to become unfashionable for half a century. It may be too strong to claim that two powerful and brilliant essays by well-respected scientists, published within a few years of each other in the 1920s, re-introduced the quest for the origin of life to respectable scientific circles, but they are certainly symbolic of a resurgent interest. Both are entitled 'The Origin of Life'. The first was written in 1922 by the Russian biochemist Alexander Oparin, but wasn't translated into English until 1967. The second was written by the maverick, self-experimenting biologist J. B. S. Haldane and published in the *Rationalist Annual* in 1929. It's always difficult to choose adjectives to describe Haldane; perhaps it's best to say 'brilliant' and leave it at that. My favourite quote of his concerns a perforated ear-drum, which he inflicted upon himself in a decompression chamber whilst trying to investigate the effects of varying oxygen levels on the human body: 'The drum generally heals up; and if a hole remains in it, although one is somewhat deaf, one can blow tobacco smoke out of the ear in question, which is a social accomplishment.'

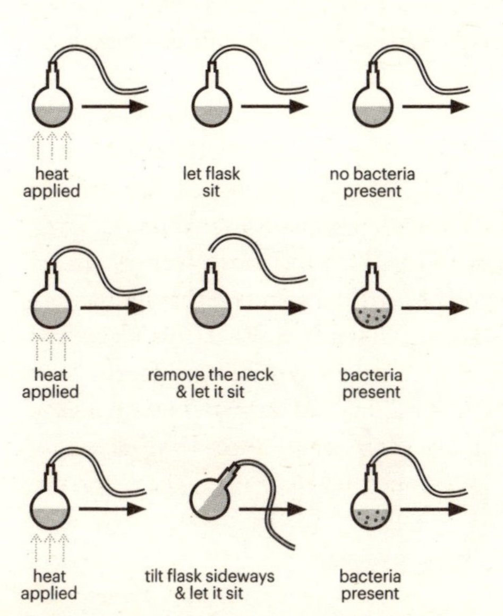

Left: Louis Pasteur's pasteurisation experiment illustrates the fact that the spoilage of liquid was caused by particles in the air rather than the air itself. These experiments were important pieces of evidence supporting the Germ Theory of Disease.

Below: Stanley Miller's apparatus enabled a glass flask containing methane, ammonia and hydrogen to simulate the reducing atmosphere of Earth and a flask of heated water which created vapour, and a pair of electrodes to mimic the presence of lightning. The 'primordial soup' was then delivered into the closed system where it was cooled and condensed into the trap at the bottom.

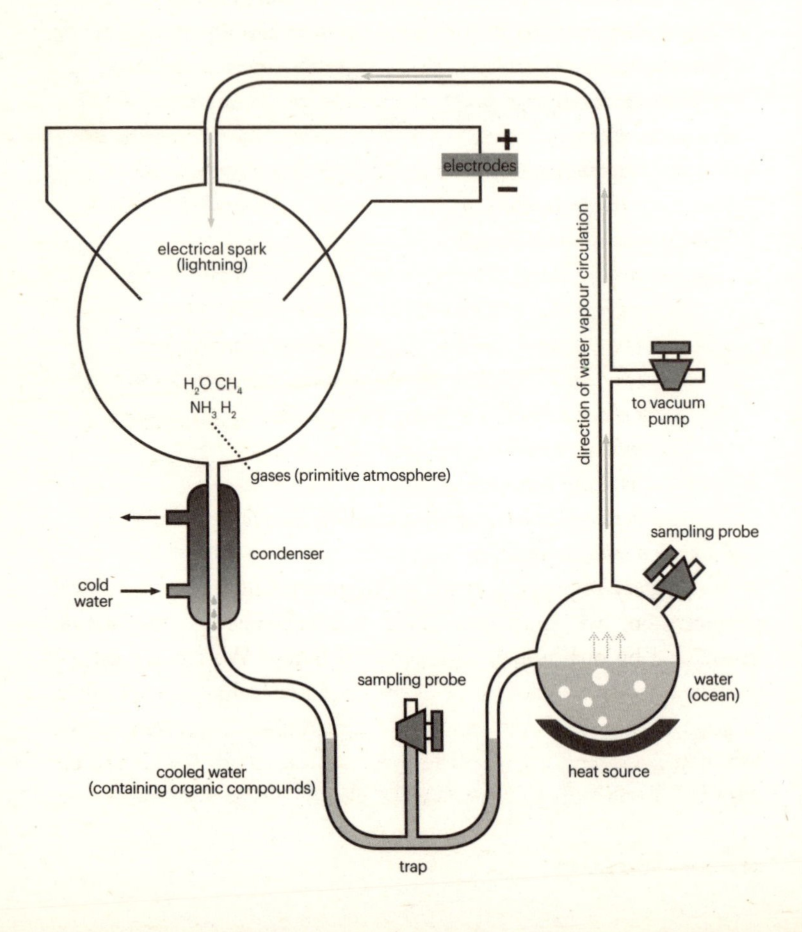

Neither scientist was aware of the other's work, but they reached similar conclusions in their eloquently argued essays. Both begin by stating the obvious question raised by Pasteur's assertion that life can arise only from life. Oparin writes:

'Pasteur's experiments showed beyond doubt that the spontaneous generation of microbes in organic infusion does occur. All living organisms develop from germs, that is to say, they owe their origins to other living things. But how did the first living things arise? How did life originate on Earth?'

The idea that life could have its origin beyond Earth is raised by both authors, and set aside. It may be correct, as we have already discussed, but it's not a useful working hypothesis because, as Oparin notes, it 'is only the answer to the problem of the origin of earthly life and not in any way to that of the origin of life in general'.

Oparin then turns to the difference between biology and chemistry:

'Do we have any logical right to accept the fundamental difference between the living and the dead? Are there any facts in the world around us which convince us that life has existed for ever and that it has so little in common with dead matter that it could never, under any circumstances, have been formed or derived from it?'

His answer is an unequivocal no.

'The specific peculiarity of living organisms is only that in them there have been collected and integrated an extremely complicated combination of a large number of properties and characteristics which are present in isolation in various dead, inorganic bodies. Life is not characterised by any special properties but by a definite, specific combination of these properties.'

Haldane is more succinct:

'The link between living and dead matter is therefore somewhere between a cell and an atom.'

This is very important. If we are to understand life as a physical phenomenon, we must put aside the extraneous complication introduced by our human experience of living. We are not asking questions about consciousness, or the origin of feelings, or morality, or good or evil, or the other infinite complexities generated *by* life. We should focus only on the difference between an atom and a single cell, and under what circumstances atoms can self-assemble into structures that we would recognise as being alive.

Haldane's and Oparin's essays are lessons in how to think carefully about a difficult problem, and it is remarkable how closely their speculations foreshadow current ideas on the origin of life, especially given the limited understanding of biochemistry available to them. The details of reparation and photosynthesis were sketchy at best, and the discovery of DNA was a scientific lifetime away. Both essays suggest the most probable location for the origin of life as a 'primeval' or 'primitive' ocean where, in Oparin's words, 'individual components of organic substances floating in the water met and combined with one another' until, switching to Haldane, it 'reached the consistency of hot dilute soup'.

The idea of a 'prebiotic soup', Darwin's warm little pond, supporting the gradual development of ever more complex organic chemistry, energised by ultraviolet light and a reactive atmosphere, is perhaps the most common picture of the origin of life in popular culture today. This is in part due to a famous experiment carried out in 1953 by Nobel Prize-winning chemist Harold Urey and his PhD student Stanley Miller at the University of Chicago. It's perhaps not surprising that the Urey–Miller experiment immediately captured the public imagination, comfortably eclipsing Crick and Watson's discovery of the structure of DNA that same year. Haldane closed his essay by creating a vivid and compelling picture of what they were attempting: 'The above conclusions are speculative. They will remain so until living creatures have been synthesized in the biochemical laboratory. We are a long way from that goal.'

Urey and Miller constructed a model primeval ocean inside a 5-litre sterilised glass flask filled with methane, ammonia and hydrogen to simulate the highly reactive reducing atmosphere that was thought to have existed on the young Earth. A pair of electrodes sent continuous sparks into the flask, mimicking the presence of lightning. The resulting 'soup' was then delivered into a cooler flask, the ancient ocean, from which samples could be extracted. The apparatus is shown on page 160.

After a single day, the primeval ocean in the flask turned an intriguing shade of pink. The experiment ran continuously for just over a week, at which point the ocean in the sterilised flask was tested for signs of organic life. Urey and Miller found amino acids, the building blocks of proteins, the basic components of life. The public response to the experiment was one of great excitement; Miller appeared on the front of *Time* magazine in 1953, whilst Crick and Watson had to make do with the less glamorous pages of *Nature*. It's easy to see why. The Urey–Miller experiment had all the hallmarks of a microbial *Frankenstein*; the fundamental building blocks of life created from lifeless atoms by a vital spark of electricity. Perhaps if the soup were left for long enough something would crawl out.

Sixty years on, the Urey–Miller experiment still casts a long shadow over the search for the origin of life. The imagery of the shadow is probably appropriate, because the basic premise of the Urey–Miller experiment is probably wrong. The evidence from the zircons informs us that Earth's primordial atmosphere was not a reactive chemical cocktail of ammonia, methane and hydrogen. On top of that, the idea that a mixture of amino acids, gently prodded by ultraviolet light and lightning would, over millions of years, coalesce into something as complex as a living cell is highly unlikely. As Nick Lane puts it in his superb book, *Life Ascending*, if you take a sterilised tin of soup from your shelf and leave it alone for a few million years, perhaps zapping it occasionally with electricity, all that will happen is that the constituent molecules will break apart. It is not very likely that something *more complicated* than the original constituents will appear. The problem is one of physics, or to be more precise, the branch of physics known as thermodynamics.

Life, thermodynamics and entropy

Even the simplest living cell is an intricate, highly ordered structure. The smallest living things on Earth are bacteria of the genus *Mycoplasma*. They are only two ten-thousandths of a millimetre across, which still makes them over a billion times the volume of a carbon atom. The simplest known living cells in terms of the number of basic biological building blocks are symbiotic bacteria known as *Carsonella ruddii*, which contain only 182 different proteins. This isn't many, given what they have to do, which is to replicate, amongst other things, but they are still extremely complex objects made up of billions of individual interacting atoms.

The problem with the primordial soup hypothesis is that something as complex as a single cell will not emerge by chance in an isolated, gently stewing pond, no matter how long you wait. The physics behind this assertion is encoded into one of the fundamental laws of Nature, known as the second law of thermodynamics. It states that things become more disordered as time passes. A broken egg never reassembles. A dead bird decays. I've lost count of the number of times it's been pointed out to me that a song I was involved in producing many years ago called 'Things Can Only Get Better' runs counter to the second law of thermodynamics. I accept that this is the case. Things Can Only Get Worse, all things considered.

The second law of thermodynamics is often stated in the following form: The entropy of an isolated system never decreases. Roughly speaking, entropy can be thought of as a measure of how many ways the component parts of something can be arranged such that it looks the same, and this is a measure of how ordered the thing is. Higher entropy means more disordered, while lower entropy means more ordered. Living things are very highly ordered. The Austrian physicist

Ludwig Boltzmann formulated this definition, and the expression for calculating the entropy of a system in this way is written on his grave in Vienna:

$$S = k_B \, ln \, W$$

S is the entropy, *W* is the number of ways of arranging the components such that they give rise to the same outcome, k_B is a constant of proportionality known as Boltzmann's constant, and the symbol *ln* stands for natural logarithm. A higher entropy configuration corresponds to lots of ways of arranging things; a lower entropy configuration corresponds to fewer ways of arranging things.

An example might make this clearer. Think about the molecules of air in a room, all whizzing around and bumping into each other. Each molecule moves around the room at random, and could end up anywhere with equal probability, given enough time. It is very unlikely that all the molecules will end up in one corner by chance, leaving the rest of the room as a perfect vacuum. Why is this so? The answer is one of simple statistics. Allow me to introduce two little pieces of jargon, because it makes everything a lot clearer and easier to write about. This is the only excuse for jargon.

Each unique configuration of molecules in the room is known as a microstate of the system. If we want to describe a particular microstate, we need to know the positions and velocities of every single air molecule. We might decide, quite rightly, that this is not something we're particularly interested in. We're more interested in things we can observe, like the temperature and air pressure distributions in the room. This more coarsely defined, but more practical characterisation of the state of the room is known as a macrostate.

If each particular configuration of air molecules – each microstate – is equally likely to occur,[2] then it follows that the room will be more likely to be in the macrostate that corresponds to the largest number of microstates. Even if we started out with all the molecules in the corner, over time they would end up filling the room. Our system will always head towards the macrostate that consists of the highest number of microstates, which is to say that it will always increase its entropy. The *W* in Boltzmann's formula is the number of microstates corresponding to a given macrostate.

This is the content of the second law of thermodynamics, and it's hard to argue with it, which is why the physicist Sir Arthur Eddington once said,

> *'If someone points out to you that your pet theory of the universe is in disagreement with Maxwell's equations – then so much the worse for Maxwell's equations. If it is found to be contradicted by observation – well, these experimentalists do bungle things sometimes. But if your theory is found to be against the second law of thermodynamics I can give you no hope; there is nothing for it but to collapse in deepest humiliation.'*

Life appears to run counter to the second law of thermodynamics, because living things are highly ordered. They are macrostates that correspond to a very few microstates, and therefore have a very low entropy. 'Did I request thee, Maker, from my clay to mould me man?' An 80kg lump of clay may have all the ingredients necessary to build a human being (it doesn't, but this is a metaphor), but most random reconfigurations of the ingredients will result in differently configured but indistinguishable lumps of clay. We'd be surprised if we got lucky and arrived by chance in the very particular configuration of ingredients that can sit up and start considering the origin of life. A human seems to be a gross violation of statistical common sense, a physicist doubly so, although I've been called worse. A bacterium is not much better. That's also been said. We've taken a little literary latitude here to make a point, however. As we have already noted, we shouldn't get confused by trying to explain how an organism as complex as a human being emerged from some sort of primordial clay 'in one go', because evolution by natural selection does most of the work. Natural selection is a non-random process and one that can drive increases in the complexity of living things quite astoundingly quickly. Having said that, evolution by natural selection has to get going in the first place, and this certainly requires some form of genetic code that can pass information down the generations, as well as all the associated proteins and machinery needed for the copying and replication of genes. We do seem to have a problem.

One of the first scientists to think carefully about this apparent paradox and to offer a solution was Erwin Schrödinger, who is best known for his foundational work in quantum theory. In 1943, Schrödinger gave a series of lectures at Trinity College, Dublin, in which he posed the question: 'How can the events in space and time, which take place within the spatial boundary of a living organism, be accounted for by physics and chemistry?' The answer, as Schrödinger noted, is that the events within the boundary of an organism cannot be understood in isolation, because organisms are not isolated systems. They can be understood only when viewed as intimately and essentially coupled to their external environment. If I am allowed two literary allusions in a single sentence without performing the statistically unlikely feat of transforming into Morrissey, I might counter Milton with John Donne; a maker is not required to mould a man because no man is an island.

If you take the 7×10^{27} atoms that make up the average human – mainly oxygen, carbon, hydrogen, nitrogen, calcium, phosphorus, potassium, sulphur, sodium, chlorine and magnesium – and throw them into a box, the result will be a high-entropy uniform distribution of atoms, just like the air molecules spread uniformly about a room. It will be very difficult to encourage them all to 'get into the tiny corner' that corresponds to a human being. You might be able to encourage some of the atomic ingredients to form structures by throwing a match into the box, however; there would be a bang as hydrogen and oxygen bind together to form water, but you'd be rightly surprised if a man emerged.

And yet a molecule of water is a lower entropy arrangement of two hydrogen atoms and an oxygen atom than would be the case if they weren't bound together, and that appears to run contrary to the second law of thermodynamics. What has happened here? The answer is very important. Whilst the entropy of the system of atoms has been lowered by the chemical reaction, a large amount of heat was released. In the jargon, an exothermic reaction has taken place. It went bang. This heat is absorbed by the surroundings, increasing the entropy of the environment by more than the entropy decrease associated with the formation of the water molecules. The entropy of the entire system increases, in accord with the second law.

Living things work in the same way from a thermodynamic perspective. They can become more ordered as long as they pay their debt by exporting disorder in the form of heat into the Universe. You are exporting disorder now as you read this book. You are hastening the demise of everything that exists, bringing forward by your very existence the arrival of the time known as the heat death, when all stars have died, all black holes have evaporated away and the entirety of creation is a uniform bath of photons incapable of storing a single bit of information about the glorious adolescence of our wonderful Universe. You are doing this by burning food in oxygen from the air. This is an exothermic reaction, generating plenty of heat for export and more than compensating for the temporary low entropy configuration of your wasteful, highly ordered body. I do seem to be turning into Morrissey: What are the odds?

The Earth's oxygen atmosphere is absolutely necessary for the heat-generating, entropy-exporting reactions that allow us to maintain our complex structure, and this is a crucial insight. We appear to 'defy' the second law of thermodynamics, but we do not because we are not isolated systems. We are part of a larger, out-of-equilibrium system. The oxygen atmosphere is unstable and ready to react with pretty much anything, given a little nudge. We exploit this imbalance to create and maintain our highly ordered structure, and as long as we keep breathing we can do so, at the expense of radiating a large amount of heat. We are like little waterwheels, exploiting a waterfall to power our internal factories. If the waterfall dries up, the wheel stops and the factory falls to bits.

The unstable oxygen atmosphere is constantly replenished by photosynthesis, itself a biological marvel that we will investigate in some detail in Chapter Four. Photosynthesis is a remarkable process from a thermodynamic perspective. Plants and algae build complex sugars from carbon dioxide and water, decreasing the local entropy and releasing highly reactive oxygen and heat into the atmosphere in the process. How is this possible? Because of the presence of a waterfall – in this case, the temperature gradient between the surface of the Earth and the Sun. Photosynthesis, which sits at the base of the entire food chain on Earth today, is possible only because there is a great external imbalance; in this case, a glowing source of photons 93 million miles away in space.

Below: A human being is primarily made up of a pile of elements, of ingredients that come from the Earth.

Number	Name	Symbol	ppm (µg/g)	ppb (atoms)
26	iron	Fe	319000	148,000,000
8	oxygen	O	297000	482,000,000
14	silicon	Si	161000	150,000,000
12	magnesium	Mg	154000	164,000,000
28	nickel	Ni	18220	8,010,000
20	calcium	Ca	17100	11,100,000
13	aluminum	Al	15900	15,300,000
16	sulphur	S	6350	5,150,000
24	chromium	Cr	4700	2,300,000
11	sodium	Na	1800	2,000,000

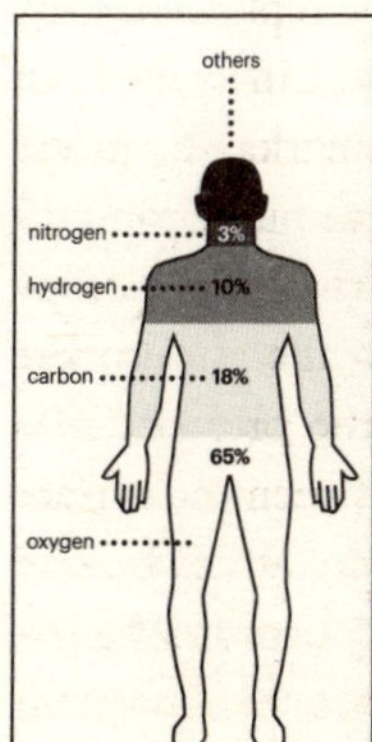

Element	Symbol	% in body
oxygen	O	65.0
carbon	C	18.5
hydrogen	H	9.5
nitrogen	N	3.2
calcium	Ca	1.5
phosphorus	P	1.0
potassium	K	0.4
sulphur	S	0.3
sodium	Na	0.2
chlorine	Cl	0.2
magnesium	Mg	0.1
Less than 0.1: Trace elements include boron (B), chromium (Cr), cobalt (Co), copper (Cu), fluorine (F), iodine (I), iron (Fe), manganese (mn), selenium (Se), silicon (Si), tin (Sn) and zinc (Zn)		

In summary, living is possible from a thermodynamic perspective because the natural environment is grossly out of equilibrium. Living things exist in the imbalances, exploiting them to build and maintain their complex structures as a mill uses a waterwheel to extract useful energy from a cascading waterfall, increasing the entropy of the entire system as it does so. In the context of the origin of life, this observation is highly suggestive. Life probably didn't begin in a gently stewing pond, because the thermodynamic gradients are too gentle to drive the emergence of complexity. Living things need to be coupled into a steady, powerful gradient from the external environment in order to build and maintain their complex structures.

The external gradients that most living things exploit today were not available to the first organisms. There was little or no oxygen in the atmosphere because photosynthesis put it there, and photosynthesis, the means by which life exploits the gradient between Sun and Earth, is an incredibly complex biochemical process that surely couldn't have predated life. The search for the origin of life therefore becomes a search for a gradient; a naturally occurring imbalance generated by Earth's geology that may have provided the spark of life; a geological cradle with a steady energy source that could drive geochemistry up the thermodynamic hill towards biochemistry.

We commented earlier in the chapter that Darwin's theory of evolution by natural selection provides the conceptual framework for the scientific exploration of the origin of life. Recall Darwin's famous lines, 'Therefore I should infer from analogy that probably all organic beings which have ever lived on this earth have descended from some one primordial form, into which life was first breathed.' This putative primordial form is a population of living things known as LUCA: the Last Universal Common Ancestor of all life on Earth.

The unbroken chain of life, stretching back 4 billion years, offers an interesting possibility. If LUCA existed, we might hope that the ensuing 4 billion years of evolution by natural selection has not removed all trace of its original biochemistry. There may be commonalities that all extant organisms share, and if so, it's likely that LUCA possessed them, too. Furthermore, if evolution, the eternal tinkerer, has not managed to replace such processes in any of the endless forms most beautiful that it has delivered during the last

4 billion years, then we might feel at liberty to conclude that these processes are fundamental and necessary components of all life. As such they would be a smoking gun, connecting living things today across 4 billion years, a third of the history of the entire Universe, to the warm little pond.

The genetic code, DNA, is one such commonality. All living things share it, from bacteria to people. There is also another, rather more surprising, thing that we all share, and that has to do with the way we manage our energy. Given what we've said about the central importance of thermodynamics to life, this is an exciting and significant observation. There is a common energy management system, and the suggestion is that this is a relic of the conditions present in the cradle of life on Earth. Living things are like books, frozen moments replete with clues about their evolutionary history. In every bacterium cell, in every blade of grass, in every cell in your body, the story of the evolution of life on Earth is documented, incompletely to be sure, but the narrative is not completely erased. Let's follow this thread to see where it leads, and explore the way that living things manage their energy.

[2] Strictly speaking we should say that this is true only in equilibrium.

The moth and the flame

At first glance, the energy-generation mechanisms employed by living things seem quite straightforward. Let us for the moment focus on animals. We burn food in air to release energy, carbon dioxide and water. The basic chemical reaction is shown below. Glucose reacts with oxygen to form carbon dioxide and water, with the release of energy. This is known as an oxidation reaction. As we discussed in Chapter One, oxygen atoms are rather keen on acquiring electrons, and will do so if they are given the opportunity. The 'burning' of sugar can be thought of as sugar molecules transferring electrons to oxygen molecules; the sugar is 'oxidised', and the oxygen is 'reduced'. If you remember anything about school chemistry, you'll probably remember 'redox' reactions, and this is an archetypal example. Redox reactions are all about the transfer of electrons, and so is life.

$$\text{Glucose} + \text{Oxygen} \rightarrow \text{Carbon dioxide} + \text{Water}$$

$$C_6H_{12}O_6 + 6O_2 \rightarrow 6CO_2 + 6H_2O$$

There are rare occasions when the necessity to find a visual texture for a television programme delivers more than wallpaper. This is one such occasion. The story of the origin of life, perhaps inevitably, has a gothic tinge. I'm not sure whether this comes entirely from *Frankenstein* or whether there is something innately unsettling about the subject that leads inexorably into the shadows. Even Genesis is quite Bauhaus at the beginning: 'And the Earth was without form, and void; and darkness was upon the face of the deep', although it turns a bit Hendrix when the lights come on and everything is told to get fruitful and multiply: 'Behold, I have given you every herb …'

The title of this chapter comes from a visual metaphor we used during filming that goes to the heart of one of our central questions. What is the difference between living and inanimate matter? What is the difference between a moth and a flame? The basic chemical reaction that powers a moth is the oxidation of glucose. The chemical reaction that powers a candle flame is an oxidation reaction of precisely the same type, as shown below.

$$2C_{18}H_{38}\ (s) + 55\ O_2\ (g) \rightarrow 36\ CO_2 + 38\ H_2O$$

In both cases, electrons are transferred from a long-chain carbon molecule and onto oxygen, but in the case of respiration, some of the energy released in the chemical reaction is syphoned off and used to live. The process by which this happens is intricate, to say the least. In a living thing, the electron doesn't just jump straight onto the oxygen, releasing all the energy at once. That would be a flame. Instead, the electron is passed between a series of atoms – usually iron – embedded in proteins that tune their appetite for electrons. There is nothing uniquely biological about iron atoms transferring their electrons to oxygen; it is known as rusting. The clever thing is the way that biological structures tune the chemistry by embedding the iron atoms in complex molecular structures, enabling them to control the flow of electrons and harness them to do useful things. This chain of embedded iron atoms, which contains around 15 steps in most organisms, is known as the respiratory chain, and it is used not only in respiration in animals, but also in photosynthesis. In one form or another, it is common to all life, and therefore certainly very ancient. All life uses redox chemistry to extract electrons from something and transport them onto something else via respiratory chains.

There is another component to the energy-management system of life that is even more intricate, and also universal. All living things store part of the energy delivered down the respiratory chain by the flow of electrons in molecules known as adenosine triphosphate, or ATP. These molecules are the universal batteries of life, transferring stored energy around your body and releasing it as needed. The way ATP molecules are manufactured is, to say the least, odd, complicated and, to be honest, downright weird. One of the great joys of making television documentaries is that I get to learn about science outside my field. I still recall how I felt when I read for the first time about how cells manufacture ATP; it was like learning that the carbon atoms in my body were manufactured in the cores of long-dead stars. It is such a wonderful story that it seems it can't be true. And, just as the fact that we are all made of starstuff connects us to the great spatial sweep of the Universe, so the story of the manufacture of ATP connects us to the great temporal sweep of the history of life on Earth. It points us back all the way to the warm little pond. Here it is.

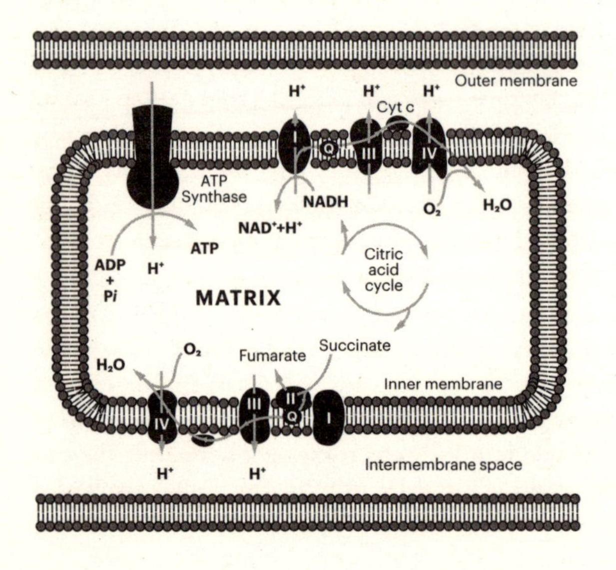

As the electrons are passed down the respiratory chain they are used to pump protons across membranes. For every pair of electrons that makes its way through, ten protons are pumped. The proton gradients are huge. In the vicinity of the membranes, which are only 6 billionths of a metre thick, the electric field strength is 30 million volts per metre, which is roughly what you'd experience if you got hit by a bolt of lightning. This great reservoir of proton potential is used to power a machine known as ATP Synthase, a nano-factory that mints new ATP molecules out of two 'empty' molecular battery components known as ADP and P_i. The protons cascade from their reservoirs down great waterfalls, spinning the waterwheel of the ATP Synthase machine at over 100 revolutions per second. The illustration, left, shows a picture of this exquisite biological machine, shared, along with DNA, by every living thing on the planet.

The intricate chemistry and structure of the respiratory chain, and in particular the use of the proton waterfalls through ATP Synthase to manufacture ATP, the universal battery of life, is surely telling us something about the deep history of life. Recall that we are searching for clues in the biochemistry of living organisms today that might point to the biochemistry of LUCA, and we have found one; the universal use of proton waterfalls as the energy source for the production of ATP. As we emphasised earlier, thermodynamics is key to understanding how life works, and surely how it began. Living things are the most complex physical structures we know of anywhere in the Universe, and building complexity spontaneously from simple building blocks is a delicate business. Life achieves it, in accord with the unbreakable second law of thermodynamics, by using redox reactions to pump protons around. This must be telling us something about how it got going in the first place? You'd be surprised, after all this, if the answer was no!

A very different Eden

Everything we've discussed in this chapter so far is established science. You will find it all in textbooks. We are now going to bring everything together and present a theory of the origin of life on Earth. This is still science, but science at the cutting edge. Some biologists agree with this theory and some don't, and this is as it should be when new ideas are in the process of forming and being tested. The theory may turn out to be wrong, and if so, its proponents will be delighted because they have learned something about Nature. It didn't happen this way. Real scientists are delighted when they find out they are wrong, and to me that is one of the greatest gifts that a scientific education can bring. There are too many people in this world who want to be right, and too few who just want to know.

Let's revisit the logic of the argument. We assume that life began on Earth, and we have evidence that this happened at some point earlier than 3.5 billion years ago. We know that the thermodynamic barrier to complexity is great, and we know that in order to overcome this, life must operate in an out-of-equilibrium system; it exists in a waterfall. Today, the waterfalls are the oxygen atmosphere and the Sun, via photosynthesis, and neither was available to the earliest life. There are other waterfalls hidden inside living things – the proton waterfalls that power the great ATP Synthase nano-factories – and these are *universal*; everything on Earth today, with a very few exceptions, uses protons. This suggests that we are looking at very ancient biochemistry; the biochemistry of LUCA.

Today, living things go to extraordinary lengths to create their internal proton waterfalls, using the complex machinery of the respiratory chains, but what if this is a later addition? What if the original energy source that drove life up the thermodynamic hill from geochemistry to biochemistry was a proton gradient? This leads to *the* question: Was there a place on Earth 3.5 billion years ago where naturally occurring proton gradients could have been harnessed by the first biological machines, allowing for the foundations of life to spontaneously emerge, all the way up to and including DNA, the prerequisite for evolution by natural selection? The answer is yes. What is more, such places still exist on Earth today, and we can visit them.

Hydrothermal vents are cracks in the ocean floor where freshwater heated by geothermal energy to over 300 degrees Celsius meets the cold saltwater of the sea. I visited a vent system whilst filming *Wonders of the Solar System* in 2009, 2000 metres down in the Sea of Cortez, just off Mexico's Baja Peninsula. Past the bioluminescence, beyond the Sun, the lights of the *Alvin* submarine brought a world of rock chimneys and tubeworms into view. It is an ecosystem founded upon clever bacteria that can drag electrons off volcanic hydrogen sulphide – redox reactions again – leaving residue mats of yellow sulphur across the vent fields. Vents like these are known as black smokers, after the particles they bellow out into the ocean.

In December 2000, whilst diving on the submerged mountain range known as the Atlantic Massif between Bermuda and the Canary Islands, *Alvin* discovered a different sort of vent system. There are great towers of calcium carbonate, some 60 metres high, raised by warm waters rich in minerals and reactive gases bubbling up from the deep crust. There is something of the fairytale spires about the place, which is why it was named the 'Lost City'.

The chemistry of the Lost City vents is different to that of those I visited in the Sea of Cortez. The waters in the vents are much cooler, around 90 degrees Celsius, because the vents are not volcanic in origin. Chemical reactions between warm water and the rocks of the sea floor saturate the structures with methane and hydrogen gases, rather than the volcanic hydrogen sulphide of the black smokers. The conditions are rather like those in the Urey–Miller experiment, which led to a broth of amino acids – the building blocks of life. The chemical origin of the vents makes a big difference to the pH level inside their porous rocky chambers; black smokers are acidic, whilst the Lost City's vents are alkaline. These terms may immediately be suggestive to you; acid means an excess of protons, and alkaline

LIVING THINGS ARE THE MOST COMPLEX PHYSICAL STRUCTURES WE KNOW OF ANYWHERE IN THE UNIVERSE, AND BUILDING COMPLEXITY SPONTANEOUSLY FROM SIMPLE BUILDING BLOCKS IS A DELICATE BUSINESS.

means a deficit of protons.

Four billion years ago, the oceans of our planet were acidic, which means they contained an excess of protons. This acidic seawater would have surrounded the alkaline vent systems like those at the Lost City, delivering a natural gradient of protons through the myriad chambers of the towers. The chambers themselves would have been lined with iron and nickel, present in large quantities in the primordial oceans, which act as catalysts in organic chemical reactions. Conditions were stable, warm, permeated with natural proton gradients and, with the unusual presence of hydrogen gas, highly reactive.

Could it be that this is what LUCA looked like? Not a cell, not a little thing like a bacterium or archaeon, but a warm rocky chamber in a vent system? The argument is compelling, at least to me. Life's proton gradients, which are absolutely central to the production of ATP, are a smoking gun. The presence of highly reactive hydrogen gas in the vents is another. As we'll see in the next chapter, photosynthetic organisms go to extraordinary lengths to stick protons – hydrogen – onto carbon dioxide to make sugars. This is fundamental to life, but it happens spontaneously in the presence of hydrogen. You don't need the machinery of photosynthesis if you have hydrogen around, and you don't need the respiratory chain to pump protons across membranes if you have naturally occurring proton gradients coursing through your chambers. Everything, from the reactive precursors of organic molecules, complete with catalysts, to the proton gradients to drive the climb up the entropic gradient, is present and correct.

If this theory is right, the basic machinery of life, up to and including DNA, was formed inside the rocky chambers of vent systems like those found at the Lost City, and you carry the evidence inside you to this day. Inside your cells, you are recreating the conditions that were present in the primordial oceans of Earth 4 billion years ago. You are making proton waterfalls, because that's what life has always done. When the chemistry inside the vents became sufficiently complex to begin replicating, passing genes down the generations, natural selection could begin to weave its magic. Life found a way to manufacture its own proton gradients, using the out-of-equilibrium conditions beyond the vents, and it put a bag around the whole thing and left. And that is how you came to be.

Life beyond Earth

The theory for the origin of life on Earth that we've presented here is certainly plausible, and there are a number of well-respected biologists who support it. As we've emphasised throughout the book, however, an argument from authority is no argument at all. Is there any way the idea that life began in vents could be tested? One way would be to build an artificial vent in the laboratory, in much the same way as Urey and Miller constructed an artificial warm little pond. A group run by Nick Lane at University College London is doing just this, with the aim of observing how complex organic chemistry might emerge in out-of-equilibrium conditions such as those found in the Lost City vents.

There is another possibility, though. If it is true that the spontaneous emergence of life is near-inevitable, given the right conditions, and that vent systems were the cradle of life on Earth, we might expect life to be present on any world that has alkaline vent systems in mildly acidic oceans. It is terrifically exciting that we may well have discovered at least one such world on our own doorstep.

In February 2005, NASA's Cassini spacecraft began to detect something strange about a small icy moon called Enceladus. The moon is only 310 miles across, and the Voyager spacecraft that passed through the Saturnian system in the early 1980s did not return detailed images of its surface. Cassini's precision measurements of Saturn's magnetic field showed that Enceladus appeared to have something like an atmosphere that was distorting the magnetic field of the planet in the vicinity of the moon. Cassini was sent in to have a closer look, and in the words of project scientist Linda Spilker, the discoveries 'changed the direction of planetary science'.

The photograph on page 12 of the plate section was taken by Cassini in October 2015 as it swept low over the surface of Enceladus. The spectacular plumes are made of water, erupting from the surface at 800 miles an hour. They emerge from hot spots on the surface known as the tiger stripes. The ejected material forms the majority of Saturn's outermost ring, known as the E-ring. When Cassini flew through the E-ring, it detected the presence of silica nanograins, which are formed when water interacts with rock at temperatures above 90 degrees Celsius. The plumes themselves are rich in organic molecules, including carbon dioxide, and recent analysis confirms that they are alkaline. Precision measurements of Enceladus's orbit suggest the presence of a subsurface ocean below the South Pole of the moon, perhaps 6 kilometres deep. Bringing all the evidence together, it appears that there is an active hydrothermal vent system on Enceladus, driving plumes of water, rich in organics, out into space. The search is now on for traces of hydrogen in the plumes, which would suggest even more strongly that Enceladus has all the conditions believed to be necessary for the spontaneous emergence of life.

This tiny moon, in the frozen outer reaches of the Solar System, a billion miles away, appears to possess a deep oceanic environment very similar to that on our planet 4 billion years ago. If this is correct, and if life is close to inevitable in such conditions, then might we expect to find signs of biology in the plumes of Enceladus? We must pose a question, rather than make an assertion, because there are many variables that we don't understand. How long has Enceladus been active? Could the ocean be a temporary phenomenon, driven by the details of her orbit today, which may well have been different in the past? We need a dedicated mission to Saturn to answer these questions, and if it were up to me, I'd start building the spacecraft tomorrow, because the ice plumes of Enceladus provide us with access to the chemistry, or biochemistry, of an alien subterranean ocean. We don't even have to land.

I think this is of overwhelming importance. We may never understand how the Universe began, but we are close to understanding how *we* began. This is surely one of the most profound questions of this or any age, as evidenced by the repeated incursions into the territory by philosophy and theology. But the origin of life is a scientific question, and not a metaphysical one. As Haldane wrote in 'The Origin of Life':

> *'Some people will consider it a sufficient refutation of the above theories to say that they are materialistic, and that materialism can be refuted on philosophical grounds. They are no doubt compatible with materialism, but also with other philosophical tenets. The facts are, after all, fairly plain.*
>
> *'The question at issue is: "How did the first such system on this planet originate?" This is a historical problem to which I have given a very tentative answer on the not unreasonable hypothesis that a thousand million years ago matter obeyed the same laws that it does today.'*

Almost a century on, we know much more about the historical problem than Haldane. We have precise dates for the origin of life on Earth, and a strong candidate for its incubator. We can see how geology might have become biology, and we understand how biological systems, over billions of years, can become sophisticated enough to inquire about their own origins. Can it really be true that the chemical elements, given an ocean, a vent and 4 billion years, can come to understand themselves? I think we are close to an answer.

Then again, since this chapter has a somewhat gothic feel, I think it should end with a hollow laugh, or perhaps the music from Roald Dahl's *Tales of the Unexpected*. Let me leave you with a memory I have of a short story by Arthur C. Clarke called 'The Nine Billion Names of God'. In it, the monks at a Tibetan monastery commission a giant supercomputer to compile a list of all the possible names of God. This, the monks suggest, is the purpose for which the human race was created. They are to know their creator in every detail. The engineers, with a wry smile, sell the monks the computer, install it, and leave the monastery after dusk to head down the mountain. Should be finished about now, says one of the engineers, but his colleague is silent. 'Overhead, without any fuss, the stars were going out.'

CAN IT REALLY BE TRUE THAT THE CHEMICAL ELEMENTS, GIVEN AN OCEAN, A VENT, AND 4 BILLION YEARS, CAN COME TO UNDERSTAND THEMSELVES?

COL

O U R

Pale Blue Dot

There is a picture of us all; a point of light in the dark. This is the Earth, viewed from the Voyager 1 spacecraft on 14 February 1990 from a distance of 6 billion kilometres. The radio waves carrying the image took five and a half hours to make the journey home.

The Voyager missions were launched in 1977 towards the gas giant planets Jupiter and Saturn, with the possibility of a continuation outwards to Uranus and Neptune afforded by a once-in-a-175-year planetary alignment. Voyager 2 reached Neptune in the summer of 1989. The spacecraft's parting glance at the frozen blue planet and her moon, Triton, is one of my favourite photographs. Delicate crescents in the dark at minus 240 degrees Celsius. Cold silence unseen for 4.5 billion years, captured once by a car-sized explorer from Earth. I have no idea when, if ever, this view will be enjoyed again.

Voyager 1 took a different path, flying close above the cloud tops of Titan, Saturn's giant moon, on 12 November 1980. The flyby catapulted the spacecraft upwards out of the plane of the Solar System on a journey into interstellar space. For a decade, Voyager 1 flew away from the Sun at a speed of 17 kilometres per second, until Carl Sagan persuaded NASA to swing the spacecraft's cameras around one last time to take a family portrait; a farewell snapshot of her home solar system as she left for the stars. Thirty-two degrees above the ecliptic plane, Voyager 1 returned a mosaic of sixty frames. Neptune, Uranus, Saturn, Jupiter, Venus and Earth are all there; only Mercury and Mars were unseen. Earth is a crescent, a tenth of a pixel in size, suspended by pure coincidence in an ochre ray from the Sun scattered in the camera's optical system.

Carl Sagan named this photograph 'Pale Blue Dot' and turned it

‘OUR PLANET IS A LONELY SPECK IN THE GREAT ENVELOPING COSMIC DARK. IN OUR OBSCURITY, IN ALL THIS VASTNESS, THERE IS NO HINT THAT HELP WILL COME FROM ELSEWHERE TO SAVE US FROM OURSELVES.’

– CARL SAGAN

'WHAT BEAUTY. I SAW CLOUDS AND THEIR LIGHT SHADOWS ON THE DISTANT DEAR EARTH... THE WATER LOOKED LIKE DARKISH, SLIGHTLY GLEAMING SPOTS... WHEN I WATCHED THE HORIZON, I SAW THE ABRUPT, CONTRASTING TRANSITION FROM THE EARTH'S LIGHT-COLOURED SURFACE TO THE ABSOLUTELY BLACK SKY. I ENJOYED THE RICH COLOUR SPECTRUM OF THE EARTH. IT IS SURROUNDED BY A LIGHT BLUE AUREOLE THAT GRADUALLY DARKENS, BECOMING TURQUOISE, DARK BLUE, VIOLET AND FINALLY COAL BLACK.'

— YURI GAGARIN

into one of the most valuable images in history, with a powerful piece of writing of the same name.

'Our planet is a lonely speck in the great enveloping cosmic dark. In our obscurity, in all this vastness, there is no hint that help will come from elsewhere to save us from ourselves.
The Earth is the only world known so far to harbor life. There is nowhere else, at least in the near future, to which our species could migrate. Visit, yes. Settle, not yet. Like it or not, for the moment the Earth is where we make our stand.
It has been said that astronomy is a humbling and character-building experience. There is perhaps no better demonstration of the folly of human conceits than this distant image of our tiny world. To me, it underscores our responsibility to deal more kindly with one another, and to preserve and cherish the pale blue dot, the only home we've ever known.'
Carl Sagan

'As we begin to comprehend that the Earth itself is a kind of manned spaceship hurtling through the infinity of space – it will seem increasingly absurd that we have not better organized the life of the human family.'
Hubert H. Humphrey, Vice President of the United States, 26 September 1966

'When you're finally up at the Moon looking back on Earth, all those differences and nationalistic traits are pretty well going to blend, and you're going to get a concept that maybe this really is one world and why the hell can't we learn to live together like decent people.'
Frank Borman, Apollo 8, *Newsweek* magazine, 23 December 1968

If you are the sort of person who likes to overcomplicate things – perhaps the abrasive weathering of accumulated disappointment has exposed your banded cynical formations? – then you may find this naïve. I have had my share of weathering, but I think Sagan's observations are unchallengeable. Our planet is vanishingly small in the vastness, which implies that each of us is also vanishingly small.

We must come to terms with being of no cosmic significance, and this means jettisoning our personal and collective egos and valuing what we have. We can no longer assume the platform of gods, or dream of a unique place in their hearts. Science has forced us to look fixedly into an infinite universe, and its volume dilutes special pleading to a vanishingly small and pathetic whimper. And yet what's left is better. No monument to the gods is as magnificent as the story of our planet; of the origin and evolution of life on the rare Earth and the rise of a fledgling civilisation taking its first steps into the dark. We stand related to every one of Darwin's endless, most beautiful forms, each of us connected at some branch in the unbroken chain of life stretching back 4 billion years. We share more in common with bacteria than we do with any living things out there amongst the stars, should they exist, and they are more worthy of our attention. Build cathedrals in praise of bacteria; we are on our own, and as the dominant intellect we are responsible for our planet in its magnificent and fragile entirety.

If this sounds hopelessly hippy, ask yourself who else might be considered responsible? Sagan is right, astronomy is humbling, and humility is the first step towards forging a better and a more secure future. Voyager's gift to its creators, delivered in a final glance, is humility, from which responsibility follows. Humility, awe and curiosity in the face of Nature are the province of science. We must accept that science has forced us to grow up, and that is a rich and fulfilling position for the human race to find itself in.

The planets are dots in Voyager's mosaic, but they are not featureless. Their shapes and sizes may be beyond the resolution of Voyager's 1970s tube television camera, but there is information in the few photons that made it through the lens. If there are living things out there beyond the two Voyagers, sophisticated and enlightened enough to do science, what could they make of our Pale Blue Dot? They would have only light, but light can carry information across interstellar distances if you know how to decode the messages it contains. The colours of a world are an encrypted database carrying the fingerprints of its constituents and chemistry. Understanding the physical nature of light and the mechanisms by which it is emitted and absorbed allows for the information to be extracted.

In this chapter we'll explore what we know about light and its interaction with matter, and how astronomers are taking the first steps to search for life beyond the Solar System by studying the light reflected and absorbed by the pale dots around distant stars. We'll also follow a parallel path; the study of light, motivated by curiosity alone, has led to discoveries over a thousand years that are both useful in a utilitarian sense and fascinating on a purely intellectual level. The simplest questions about the origin and nature of light and its interaction with matter are still delivering exotic answers today at the edge of our knowledge, and this is precisely where we should be. If, together, we can learn to gaze at the lights of the night sky with excitement and joy and curiosity and with no fear of the infinite unknown, we will have chosen a future free of superstition, driven by the quest for an ever-deeper understanding of Nature, freed from the shackles of absolute certainty, save for the recognition of our absolute responsibility for our planet and ourselves.

The rainbow connection

Where do the colours of the world come from? The Earth has no light of its own, at least, not if we neglect the electric glow of our civilisation. The Pale Blue Dot is a reflector, a mote of dust catching the rays of the Sun. The Sun's light isn't inherently blue; it contains all the colours of the rainbow. Of course it does, it's one of the first things we learn at school. Yet common knowledge is often hard-won. The early development of our scientific understanding of light is intertwined with the question of the origin of rainbows, and this doesn't surprise me.

Rainbows are amongst Nature's most intriguing regular forms, a glorious arc on elemental days. They appear above all landscapes on Earth, at any time from dawn until dusk, and yet all share a set of universal properties: the colours always appear in the same order, no matter what the weather. Red points skywards, blue to the ground. They are always centred on the observer, a personal universal phenomenon, and all arc across the sky subtending the same angle between the bright rays of the Sun and the eye of the observer: 42 degrees. A bridge to heaven or a covenant from the gods, such a magnificent symbol demands a divine explanation. And if we allow ourselves for one last time to define the divine as the underlying laws of Nature, then the rainbow is one of the most vivid shadows of the deeper structures that govern the Universe; a visual representation of the behaviour of light. This is why many of the scientific greats have tried to understand them.

As far back as the eleventh century, Ibn al-Haytham searched for a physical explanation for rainbows. He correctly surmised that they are caused by light from the Sun interacting with water in the atmosphere before entering the eye – although he was incorrect in that he thought the rainbow was caused by reflections off clouds, which he believed behaved as giant concave mirrors. The suggestion that the rainbow is reflected sunlight is not a trivial observation. The theory that vision is active in the sense that the eye generates the light that allows the viewer to perceive objects, rather like a radar system, was widespread in the eleventh century, and had the historical authority of Euclid and Ptolemy to support it. Ibn al-Haytham had little time for the authority of the ancients, however, and placed great emphasis on experimentation and observation rather than pure thought and instinct. This approach, which we now recognise as distinctly modern, is one of the reasons why he is regarded by many historians of science as one of the great early scientific minds. As the historian David C. Lindberg writes, he was 'undoubtedly the most significant figure in the history of optics between antiquity and the seventeenth century'.

Alhazen (the Latinised version of his name) is not as well known as Newton, Galileo, Kepler or Einstein, but I think he deserves a much more prominent place in the history of science because of the self-awareness and humility that is evident in his writings; essential components of the modern scientific enterprise which echo Sagan's thoughts on the Pale Blue Dot. All good research scientists understand that no position is unassailable; there are no absolute truths in science; authority counts for nothing when contradicted by Nature; *nullius in verba*. Here is Alhazen, writing in Basra a thousand years ago:

> *'Therefore, the seeker after the truth is not one who studies the writings of the ancients and, following his natural disposition, puts his trust in them, but rather the one who suspects his faith in them and questions what he gathers from them, the one who submits to argument and demonstration, and not to the sayings of a human being whose nature is fraught with all kinds of imperfection and deficiency. Thus the duty of the man who investigates the writings of scientists, if learning the truth is his goal, is to make himself an enemy of all that he reads, and, applying his mind to the core and margins of its content, attack it from every side. He should also suspect himself as he performs his critical examination of it, so that he may avoid falling into either prejudice or leniency.'*

Alhazen's greatest surviving work (half his writings have been lost), *The Book of Optics*, was the inspiration for many of the subsequent investigations into the origin of the rainbow and the nature of light. There is no irony here; books are to be read critically. They are not sources of 'truth', but of inspiration; snapshots of knowledge and experience which should be read with a critical eye. It is a measure of the power of the written word that Alhazen's book inspired generations of scientists from cultures widely separated in space and time to seek to improve on his work, as he implored them to do, and to find a rational and experimentally testable explanation for the rainbow.

Kamal al-Din al-Farisi was one of a long line of pioneering scientists who created a vibrant academic culture throughout Persia during the late medieval period. Born in 1265, al-Farisi completed his studies under the tutelage of astronomer Qutb al-Din al-Shirazi at the celebrated Maragheh Observatory near Maragheh, Iran. Al-Farisi became interested in the refraction of light – the bending of light rays when they pass from air into water or glass. Al-Shirazi told al-Farisi to read *The Book of Optics*, and he became so engrossed in it that al-Shirazi encouraged him to write an updated review of its contents. The result was a complete revision of the work, and a step towards a correct explanation for the formation of rainbows. Al-Farisi suggested that a rainbow is formed by light entering water droplets from the air, being refracted twice – once on entering and once on leaving the drop – and undergoing at least one reflection from the back surface. Following Alhazen's eloquent entreaties, he conducted a series of experiments to test his theoretical approach; a beautiful early example of the controlled exploration of Nature under laboratory conditions. Al-Farisi created a model of a rain-laden atmosphere using large spherical glass vessels filled with water. He placed his glass raindrops into the equivalent of a camera obscura, a dark room with a controlled aperture through which to introduce a beam of sunlight, and flat surface on which to project an image. He observed a rainbow, verifying the broad outline of his theory.

At virtually the same time, but widely separated geographically, the German monk and scholar Theodoric of Freiberg arrived at the same conclusion, documented in *De iride et radialibus impressionibus*

LET'S CHANT THE GLORIES OF SURYA, WHOSE BEAUTY RIVALS THAT OF A FLOWER. I BOW DOWN TO HIM, THE RADIANT SON OF SAINT KASHYAPA, THE ENEMY OF DARKNESS AND DESTROYER OF EVERY SIN.

— A PRAYER OF CHHATH PUJA

– 'On the rainbow and the impressions created by irradiance'. Just like al-Farisi, Freiberg used glass spheres filled with water to model raindrops and explored the interaction between sunlight and water. Despite being thousands of miles apart and with no contact or communication, it is not a coincidence that these two early scientists arrived at the same conclusions almost simultaneously. Both were inspired by and built upon *The Book of Optics*, which was translated from Arabic into Latin in the twelfth century and disseminated around Europe as well as Persia – an early example of an essential principle that we still fight for today; scientific knowledge must be freely available through open publication. There must be no copyright on ideas.

These were great steps forward, but neither scientist discovered the correct explanation for the origin of the rainbow's colours or their most striking geometrical property; the universal angle of the circular arc.

René Descartes was the first to explain the geometry of the rainbow in a 1637 essay entitled *L'arc en ciel*. Perhaps unsurprisingly from the father of Cartesian geometry, his method was geometrical. *L'arc en ciel* contains a well-known and beautiful diagram which marks out all the angles and lines associated with the formation of a rainbow. Here is how Descartes described the diagram.

> *'I found that if the sunlight came, for example, from the part of the sky which is marked AFZ and my eye was at the point E, when I put the globe in position BCD, its part D appeared all red, and much more brilliant than the rest of it; and that whether I approached it or receded from it, or put it on my right or my left, or even turned it round about my head, provided that the line DE always made an angle of about forty-two degrees with the line EM, which we are to think of as drawn from the centre of the sun to the eye, the part D appeared always similarly red; but that as soon as I made this angle DEM even a little larger, the red colour disappeared; and if I made the angle a little smaller, the colour did not disappear all at once, but divided itself first as if into two parts, less brilliant, and in which I could see yellow, blue, and other colours ... When I examined more particularly, in the globe BCD, what it was which made the part D appear red, I found that it was the rays of the sun which, coming from A to B, bend*

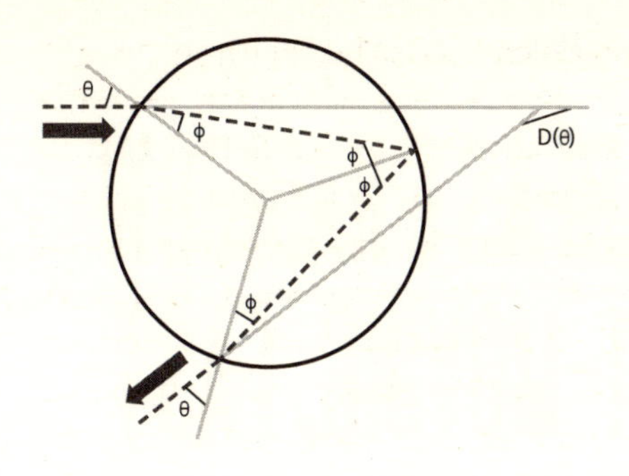

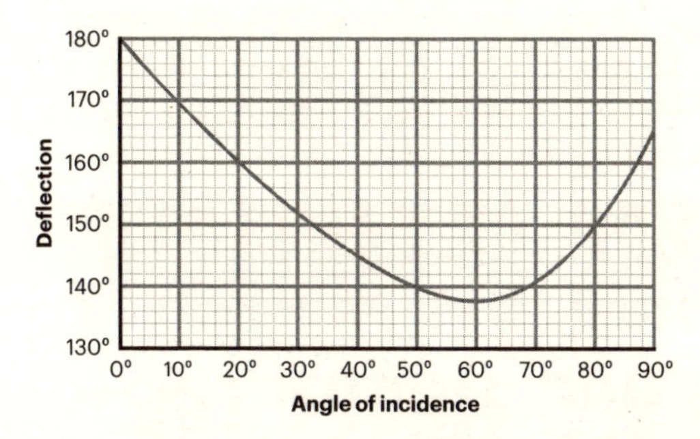

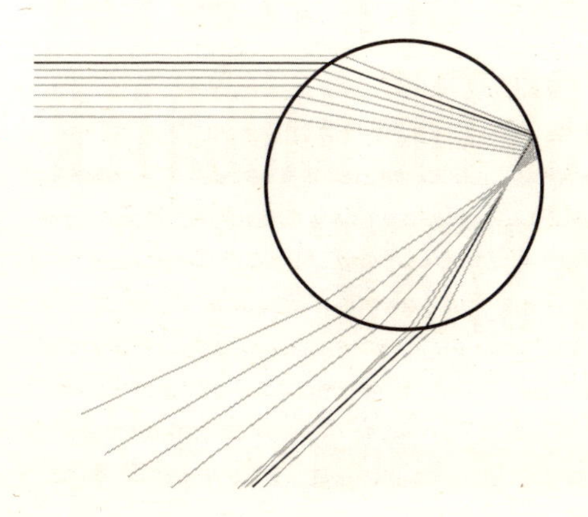

Top: A ray of light from the Sun enters a raindrop at angle θ. It is refracted into angle ϕ according to Snell's law, where $\sin\theta / \sin\phi = n_{f,water} / n_{f,air}$, and $n_{f,water}$ and $n_{f,air}$ are the refractive indices of water and air respectively. The refractive index is dependent on the colour of the light, which is the reason for the labels *f*. If you look at the diagram for quite a long time, you should be able to convince yourself that $D_f(\theta) = (\theta - \phi) + (180^\circ - 2\phi) + (\theta - \phi)$. If you want to plot $D_f(\theta)$ against θ for yourself, then substitute for ϕ using Snell's law. The refractive index $n_{f,air}$ is approximately equal to 1 for all colours, and $n_{f,water} = 1.33$ for red light and 1.34 for violet light. The outgoing angle $D_f(\theta)$ as a function of the angle of light entering the raindrop θ is shown in the illustration below.

Middle: The deflection of a ray of light entering a raindrop from the Sun at angle θ.

Bottom: A cluster of parallel rays from the Sun enter the spherical raindrop at different angles relative to the curving surface of the raindrop. On leaving the raindrop, they tend to cluster around the 'caustic' or 'rainbow' ray, shown in dark grey, which emerges at an angle of approximately 138 degrees relative to the incoming rays from the Sun. The precise angle of the rainbow ray is dependent on the colour of the incoming light, and it is this that is responsible for the colours of the rainbow.

‘THE STEADFAST RAINBOW IN THE FAST-MOVING, FAST-HURRYING HAIL-MIST. WHAT A CONGREGATION OF IMAGES AND FEELINGS, OF FANTASTIC PERMANENCE AMIDST THE RAPID CHANGE OF A TEMPEST – QUIETNESS THE DAUGHTER OF STORM.’

– SAMUEL TAYLOR COLERIDGE

on entering the water at the point B, and to pass to C, where they are reflected to D, and bending there again as they pass out of the water, proceed to the point.'

Descartes was able to explain the arc, but not the origin of the colours, because he did not know that the white light from the Sun is made up of all the colours of the rainbow. Isaac Newton made this discovery forty years later. Our purpose here is to understand the physics of the rainbow, and it is easier to explain both the geometry and the appearance of the colours at once rather than letting the story unfold chronologically, so this is what we'll do.

The top illustration on page 197 shows a ray of light from the Sun entering a raindrop, reflecting off the back surface and entering the eye of the observer of the rainbow. The angle through which the ray is deflected relative to the incoming ray is labelled $D(\theta)$. In Descartes' diagram, this is the angle between the lines AB and ED. If you're mathematically inclined, have a look at the calculation in the caption. If you don't fancy that, see the graphical representation of the angle $D(\theta)$ in the second illustration, because this is the key point. The graph shows how the angle of the light that bounces out of a raindrop changes as the angle of the light from the Sun entering the surface of the raindrop changes.

The important thing is that the angle of the outgoing light ray $D(\theta)$ has a minimum value of approximately 138 degrees. Immediately, this should ring a bell; 180 – 138 = 42 degrees, the angle Descartes came up with using geometrical methods and observation, and labelled DEM in his diagram.

To understand why this special minimum angle corresponds to a bright arc in the sky, have a look at the bottom illustration. This shows what happens to a whole bundle of rays from the Sun hitting a raindrop across a large section of its curved surface. Visually, you can see that most of the incoming rays come out at the 'special' minimum angle, even though many of them hit the raindrop at different incoming angles relative to the surface. This means that the incoming rays are preferentially focused around the minimum angle of 42 degrees, and this is why the rainbow arc across the sky is brighter than anything else. It's a focusing effect caused by the spherical geometry of raindrops. In slightly more mathematical language, rays that enter the drops over a wide range of angles clustered around 60 degrees will all emerge at a very similar angle, because $D(\theta)$ doesn't change very rapidly close to its minimum value. The special outgoing ray, corresponding to minimum deflection and shown in dark grey on the diagram, is known as the 'caustic' or 'rainbow' ray.

This explains why we see a bright arc, but not the value of the angle – 138 degrees – or the spread of colours. The value of the angle depends on how much a ray of light is refracted when it enters and leaves the raindrop; the relationship between angles θ and ϕ in the top illustration. This depends on the properties of air and water, and there is a very simple relationship between the two angles known as Snell's law, or occasionally the Snell–Descartes law. The law itself has been known to varying degrees of accuracy since classical times, and almost appears in Alhazen's *Book of Optics*, although it is not clearly stated. The history is convoluted, but the law is simple. We'll state it, and if you don't know any trigonometry then ignore it and skip to the next sentence: $\sin\theta / \sin\phi = n_{f,water} / n_{f,air}$, where $n_{f,water}$ and $n_{f,air}$ are the refractive indices of water and air respectively. The refractive index is a number that describes how light travels through a particular substance. To be specific, it describes the ratio of the speed of light in the substance to the speed of light in a vacuum. For water, the refractive index is approximately 1.3, which means light travels around 1.3 times faster in a vacuum than it does in water. For air, the refractive index is very close to 1. This is the number that sets the angle of the rainbow ray, and you can follow the calculation for yourself. In one sentence, the angle of the rainbow is 42 degrees because raindrops are made of water.

On Saturn's moon, Titan, there are raindrops of liquid methane floating delicately downwards like snowflakes in the dense atmosphere. The Sun is weak on this distant world, and the atmosphere often dominated by a thick haze, so rainbows are rare, but they will still be present if viewing conditions are right. The refractive index of liquid methane is 1.29, which leads to rainbows larger than those on Earth with an angle of 49 degrees. That's quite a big difference for a small change in the refractive index, and this is the reason for the colours of the rainbow. Even in the same substance, the refractive index of light is different for different colours. In water, the refractive index of red light is around 1.33. Blue light has a refractive index of around 1.34. When refracted through Earth's water raindrops, the rainbow made by red light is slightly larger than that made by blue light. This is why the outer ring of the rainbow is red and the inner ring is blue, with all the other colours in between. Water droplets split the white light of the Sun into the individual colours that make it up because red light travels at a slightly different speed through water than blue.

With this explanation for the origin of the rainbow, we reach the level of understanding achieved by Isaac Newton in the mid-1680s. It's not the end of the story, though – not by a long shot. There are many subtle features of rainbows that require a more advanced treatment. Double rainbows occur because there can be two reflections inside the raindrops as well as one. There can be faint arcs above and below the primary rainbow called supernumerary arcs, an interference effect first explained by Thomas Young in 1804 by treating light as a wave. The Astronomer Royal of the time, George Biddell Airy, produced a complete theory of the rainbow in 1831, which triggered mathematical research on what is known as the Airy Integral by two of the great mathematicians of the day, Augustus De Morgan and George Gabriel Stokes. This leads to more questions. If light is to be treated as a wave, what's doing the waving? Nobody knew in 1831; the answer was discovered in the early 1860s by the Scottish theoretical physicist James Clerk Maxwell – we'll get to this in more detail later – and as we've already seen in Chapter Two, Albert Einstein was the first to take Maxwell's theory of light at face value, and this led him to jettison Newtonian physics and construct a new theory of space and time. We could go on, and we will, at least down some of the paths that are opening up.

Let us pause for a moment, though, and reflect that the investigation of the rainbow is a beautiful example of the central theme of this book: simple questions about the origin of everyday things often – more often than not – lead us down tangled paths through the dense, interconnected undergrowth of physics. This shouldn't be a surprise, although it is only human to be constantly surprised at the deep interconnectedness of Nature; I would say this is one of the great joys of physics. It shouldn't be a surprise because we've established, or at least asserted with some supporting examples, that the complex world we perceive is a shadow of simpler forms: the underlying laws of Nature.

If this is the case, it must follow that there are common explanations for many of the shadows, and investigating one will inexorably lead us to touch on the deep underpinnings of another. This is why the modern trend for directing scientific research into areas deemed *a priori* economically or socially useful is not only misguided but positively harmful to the scientific enterprise, and therefore to the goal of the government advisors who dream up such daft, albeit (to be charitable) well-intentioned policies. Serendipity always was and always will be absolutely central to discovery, *because* the natural world is so intricately interconnected and functions according to a small set of fundamental laws, as far as we know. There are so many ways to discover deep and ultimately useful things that it is futile to imagine that we can predict which investigation of which tiny corner of the natural world will bear undreamt-of fruit. Nature is too complicated. Investigating rainbows might seem whimsical, but it stimulated a great deal of the early research into optics and the nature of light, and ultimately into the nature of space and time.

That said, let us continue down the particular tangled path we've chosen. The investigation of the rainbow has served as an introduction to and raised a series of questions about light that we should now seek to answer. The origin of the light that shines into the water droplets is a good place to start. As we've seen, there was vigorous debate in Alhazen's time about whether vision was an active or passive process, and the study of rainbows played a part in establishing that the light that creates them has its origin in the Sun.

Why does the Sun shine?

Why does the Sun emit light? The answer would seem to be obvious: the Sun is hot, and all hot things shine. But why do hot things shine? This is a deeper question. There is also the question of the energy source that powers the Sun and heats it up in the first place. The energy output of the Sun was well known in the nineteenth century because it is an easy thing to measure if you know the distance to the Sun. One way is to take a known volume of water with a known surface area, place it in direct sunlight and see how long it takes for the water temperature to rise by one degree. This will tell you how much energy has entered the water during the measured time. A more accurate measurement would take account of the loss of solar energy as the light travels through the atmosphere.

Observations at high altitude can help. The first measurement of the solar constant – the power output of the Sun – was made in 1838 by a Frenchman, Claude Pouillet. He estimated that around 1.2 kW of power per square metre falls on the Earth, 93 million miles away from the Sun. The modern measurement for the power delivered by the Sun at the top of the Earth's atmosphere per square metre is 1.41 kW in January, when the Earth is closest to the Sun, and 1.32 kW in July, when the Earth is furthest away; the Earth's orbit is an ellipse with the Sun at one focus. This is a colossal amount of energy. Imagine a sphere 93 million miles in radius, with enough energy to power a bright floodlight, falling on every square metre of the inside of the sphere every second. Sometimes numbers are so large that they are not helpful, but we may as well quote the total solar power output; it is 3.8×10^{23} kW. The total power-generating capacity of our civilisation today is around 16×10^{9} kW; twenty million million times smaller. Here we go again, down a tangled path suggested by a simple question. This is a vast amount of energy. What could the source possibly be?

The origin of the Sun's energy was highly controversial during the late nineteenth and early twentieth centuries, because there was no known physical process capable of sustaining such a vast energy output for more than a few thousand years, notwithstanding the enormous size and mass of our star. Again, it's very hard to picture the enormity of the Sun; a hundred Earths would line up along its diameter. It would take the average passenger jet six months to fly around it. It is traditional to say something about the size of Wales at this point; it would take 289 million countries the size of Wales to tile the surface of the Sun. But even with these vast resources of matter, the power output is difficult to explain. In 1862, Lord Kelvin, one of the greatest and most respected scientific voices of the day, declared that the Sun could be no more than 30 million years old, given its colossal power output, in direct contradiction with estimates of the age of the Earth from geological and biological evidence, which pointed to an age in excess of 300 million years.

Kelvin was over-confident and wrong, because he did not admit to the possibility of new physics providing an explanation for the source of solar energy. 'He should also suspect himself as he performs his critical examination of it, so that he may avoid falling into either prejudice or leniency'; Kelvin would have done well in this instance to read Alhazen. The new discovery was nuclear physics. Physicists like to do back-of-the-envelope calculations, and we can use one to see how nuclear physics helps. Kelvin calculated that, if the Sun were made of coal, then this vast burning repository, papered by 289 million countries the size of Wales, would contain enough 'coal' to shine with the measured brightness for 3000 years. This gives some indication of the power stored in a star. Chemical reactions such as coal burning typically proceed at energy scales a million times smaller than nuclear reactions. This is a reflection of the fact that the strong nuclear force, which binds the nucleus of atoms together, is much stronger than the electromagnetic force, which binds atoms together. Chemistry is about rearranging atoms, and nuclear physics is about rearranging nuclei. Ernest Rutherford discovered the atomic nucleus in May 1911 in Manchester, and so Kelvin knew nothing of this hidden, higher-energy layer of physics. Because nuclear reactions typically operate at energies of the order of a million times those of chemical reactions, they will increase the energy available to the Sun by a factor of around a million, give or take. This suggests an age of at least 3 billion years – much closer to the modern estimate of a 10-billion-year solar lifetime. The current best estimates of the age of the Sun from computer modelling are around 4.57 billion years, which agree nicely with the radioactive dating of meteorites in the Solar System.

The nuclear physics of the Sun

All of the four fundamental forces of Nature that we met in Chapter One are involved in the energy-releasing nuclear reactions inside the Sun's core. Stars begin their lives as clouds of hydrogen and helium, the atomic nuclei of which were formed in the first three minutes after the Big Bang. Under the action of gravity, the clouds collapse in on themselves, and the collapse heats them up. When the temperature reaches around 100,000 degrees Celsius, the hydrogen and helium nuclei can no longer hold on to their electrons and the cloud becomes a plasma – a hot gas of free electrically charged particles. As the collapse continues, the temperature rises further, and for sufficiently massive clouds the naked hydrogen nuclei approach each other with such speed that, despite their mutual electromagnetic repulsion – recall that a hydrogen nucleus is a single proton, which carries a positive electric charge – they can get very close. When this happens, a transformation takes place under the influence of the weak nuclear force. We met this transformation in passing in Chapter One as I reminisced about my time doing particle physics in Hamburg. There was a point there beyond gentle biography; we explored the constituents of matter using a particle accelerator and by consuming red wine and fine cheeses and attempting to get gout. Physics should be joyous. But I digress... Recall that a proton is made of two up quarks and one down quark and a neutron is made of two down quarks and an up quark. The weak nuclear force can turn an up quark into a down quark, so a proton can be transformed into a neutron, along with the creation of a positron and a neutrino. Neutrons are electrically neutral, and shorn of their positive electric charge carried off by the positron, they are free to approach the proton closely enough for the strong nuclear force to take over and

'DO NOT ALL FIX'D BODIES, WHEN HEATED BEYOND A CERTAIN DEGREE, EMIT LIGHT AND SHINE; AND IS NOT THIS EMISSION PERFORM'D BY THE VIBRATING MOTION OF ITS PARTS?'

bind them together tightly. The resulting atomic nucleus, made up of one proton and one neutron, is called a deuteron.

The formation of a deuteron from two protons is known as nuclear fusion. It releases a vast amount of energy because the deuteron is less massive than two free protons. Einstein discovered that mass can be transformed into energy according to the equation $E=mc^2$ (by taking Maxwell's equations describing the nature of light seriously – the tangled path), and it is the energy of fusion which is the power source of all the twinkling stars in the night sky.

The numbers involved are absolutely huge; if we could take a cubic centimetre of the Sun's interior and convert all the protons into deuterons, we could power an average-sized town for a year. Inside the Sun the fusion process doesn't stop with the creation of deuterons. Another proton quickly fuses with the deuteron to form a helium-3 nucleus, and two helium-3 nuclei then fuse together to form a helium-4 nucleus, with the release of two protons. At each stage, mass is transformed into energy, which heats up the star. This energy also halts the gravitational collapse, because the super-heated plasma exerts an outward pressure that balances the inward pull of gravity. This is why stars are long-lived structures – they exist in a delicate yet stable equilibrium, as long as they have nuclear fuel to burn in their cores. Our Sun burns six hundred million tonnes of hydrogen fuel every second into helium, with the loss of four million tonnes of mass, which is released as energy. To get a sense of how many individual fusion reactions this corresponds to, consider that there are sixty billion neutrinos per square centimetre per second passing through your head from the Sun as you read this book, and only one is released every time a proton turns into a neutron. We'll have more to say about these neutrinos later on, because they are very interesting. At this rate, the Sun has enough nuclear fuel to last another five billion years, at which point it will begin to fuse helium into carbon and oxygen before running out of options to release more fusion energy and collapsing into a fading ember known as a white dwarf.

White dwarfs are dense, exotic objects held up against the crushing force of gravity by a quantum mechanical effect known as the Pauli exclusion principle. They are planetary-sized spheres of stellar mass;

a sugar-cubed piece would weigh a tonne. The Sun's exposed carbon-oxygen core will gradually radiate its heat away, leaving a darkening ember known as a black dwarf; it will last, if not for eternity, then for a very long time. In a thousand billion years the stellar remnant will fade from view as its temperature continues to fall. Its eventual fate is dependent on physics that we have yet to understand. It is thought that matter itself is unstable over very long timescales, and if this is the case then black dwarfs will evaporate, given enough time – of which there is likely to be an infinite amount. Lower limits on the lifetime of black dwarfs suggest they should be around for at least 10^{32} years, which is ten thousand billion billion times the current age of the Universe.

Nuclear fusion is the origin of the Sun's energy, and ultimately the source of its light. The physical processes that produce the light that arrives at the Earth are different, however. It is the glowing surface of the Sun that we see in the sky, not its hidden nuclear-fired core. The surface of the Sun has a temperature of only 5500 degrees Celsius, and the light it emits is characteristic of this temperature, and not the 15 billion degrees at which the fusion reactions take place.

The idea that objects emit light according to their temperature is a familiar one. We speak of things as being 'white hot', and are familiar with the cooling red embers of a dying fire. The temperature of something is related to the colour of light it emits, and this is a clue to the origin of that light. Simple questions lead to deep answers, and the question of how hot things emit light is *the* classic example. The first thing to say is that it's an old question; Isaac Newton considered it in his treatise on light, *Opticks*, published in 1704, and his suggested answer is correct in broad outline. '*Do not all fix'd Bodies, when heated beyond a certain degree, emit Light and shine; and is not this Emission perform'd by the vibrating motion of its parts?*' It is the motion of the building blocks of matter that produces light, but it was not until the mid-nineteenth century that we began to understand the mechanism for this emission, and the quest for answers ultimately led to quantum theory and the construction of the technological foundations upon which our modern society rests.

Why do hot things shine?

Part 1: James Clerk Maxwell and the Golden Age of Wireless

Matter is constructed of electrically charged particles, and when charged particles are shaken around, they emit light. More precisely, they emit electromagnetic radiation. The discovery that light is an electromagnetic phenomenon was made by the Scottish physicist James Clerk Maxwell in a series of papers published between 1861 and 1862.

We've already met Maxwell's equations in Chapter Two, as the inspiration for Einstein's Theory of Special Relativity. To recap, Maxwell discovered a unified description of the experimental and theoretical work of a generation of physicists, including some of the great names commemorated today in the units we use to describe electricity: Volt, Ampère, Faraday, Gauss. If Maxwell's only achievement were simplification, however, Einstein wouldn't have described his work as 'the most profound and the most fruitful that physics has experienced since the time of Newton'.

Maxwell's 'profound' achievement was not merely to unify, but to discover something quite new. He discovered that light is intimately connected to electricity and magnetism in a piece of work representing one of the most vivid examples of what physicist Eugene Wigner termed the *unreasonable* effectiveness of mathematics in the physical sciences; the notion that mathematical beauty, occasionally alone, can lead to a deeper understanding of the physical world.

By the mid-1800s Faraday and others had discovered that electricity and magnetism are related. If an electric current is pulsed through a wire, a compass needle close to the wire is deflected in time with the pulse. If a magnet is moved in and out of a coil of wire, an electrical current flows through the wire whilst the magnet is moving. This is the basis of the electric motor and generator. Faraday thought deeply about the connection between the wires and the magnets. He reasoned that there must be some sort of physical link between the electrical current in a wire and the compass needle in order to deflect the needle; things don't just move of their own accord. He pictured this physical link as a 'field', which might be visualised as the pattern formed when iron filings are scattered onto a piece of paper above a magnet.

Faraday's rather mechanical idea of electric and magnetic fields was not widely accepted at the time, primarily because it didn't appear to be necessary. The mathematical equations that described electric and magnetic phenomena were written in terms of things that can be directly measured – volts and amps and forces that cause compass needle deflections. The deeper level of abstraction represented by the fields appeared to add unnecessary complication.

Maxwell discovered that this was emphatically not the case. He embraced the deeper description and rewrote all the equations describing electrical and magnetic phenomena in terms of electric and magnetic fields, rather than currents, voltages and forces. In doing so, he was forced to add an extra term into one of the equations for reasons of mathematical consistency. That term, which is called Maxwell's displacement current, had a remarkable consequence. Once present, Maxwell saw that he could rewrite his equations in a different form, known as wave equations. In this form, the equations are able to describe a self-propelling disturbance in the electric and magnetic fields.

$$\nabla^2 E - \frac{1}{c^2}\frac{\partial^2 E}{\partial t^2} = 0,$$

$$\nabla^2 B - \frac{1}{c^2}\frac{\partial^2 B}{\partial t^2} = 0.$$

Left: Maxwell's wave equations for the electric and magnetic fields.

Maxwell's wave equations can be pictured as describing energy sloshing backwards and forwards between the electric and magnetic fields, radiating outwards from an electromagnetic disturbance in the way ripples radiate outwards on a pond in response to a splashing stone. The difference is that there is no water or any other medium needed to support the disturbance – the fields themselves are sufficient to carry the energy away, one rising as the other falls. This is a fascinating observation in itself, but there was a great and most marvellous denouement. I cannot imagine how Maxwell reacted; he must have felt he was allowed a brief glimpse beyond the shadows at one of Nature's clean foundations. This self-propelling disturbance has a speed, according to Maxwell's wave equations – in the equation on page 211 it is represented by the symbol *c*. Perhaps unsurprisingly, the speed has to do with the strengths of the electric and magnetic forces – the amount by which a change in one field induces a change in the other. The speed is predicted to be the ratio of the strengths of the two forces, and Maxwell knew these quantities because Faraday and others had measured them in experiments in their laboratories. If you're familiar with a bit of electromagnetism from school, you may recognise their names and symbols; the permittivity of free space, ε_0, and the permeability of free space, μ_0. When Maxwell put the numbers in, he discovered that the speed of the disturbance came out as the speed of light! Immediately, he would have known that he had found a deeper description of the nature of light itself: light is a travelling disturbance in the electromagnetic field that drives itself along at precisely 299,792,458 metres per second.

Einstein wrote about how he imagined Maxwell must have felt in an essay entitled 'The Fundaments of Theoretical Physics': 'Imagine his feelings when the differential equations he had formulated proved to him that electromagnetic fields spread in the form of polarised waves and with the speed of light! To few men in the world has such an experience been vouchsafed. At that thrilling moment he surely never guessed that the riddling nature of light, apparently so completely solved, would continue to baffle succeeding generations. Meantime, it took physicists some decades to grasp the full significance of Maxwell's discovery, so bold was the leap that his genius forced upon the conceptions of his fellow workers.'

These words afford an insight not only into the magnitude of Maxwell's discovery, but also into the mind of a true explorer of Nature. It is amongst *the* most wonderful feelings available to a human being to understand something about the physical world for the first time. Few experience the privilege of genuine discovery, but the overwhelming excitement of understanding is available to all and is what drives a child to become a scientist.

It is interesting, and perhaps revealing of Einstein's character, that he didn't mention that Maxwell's insight turned out to be extremely useful. Heinrich Hertz confirmed the existence of Maxwell's electromagnetic waves in a series of experiments conducted between 1886 and 1889, in which he inadvertently invented the radio transmitter. I say inadvertently, because when asked by one of his students the perennial question of which scientists often tire, 'What use is all this?' Hertz replied, 'It's of no use whatsoever. This is just an experiment that proves Maestro Maxwell was right. We just have these mysterious electromagnetic waves that we cannot see with the naked eye. But they are there.'

On reading Hertz's 1888 journal article, a young Italian named Guglielmo Marconi noticed that Hertz's work could be used for signalling, and by 1901 (arguably), and certainly by 1902, he had transmitted messages using radio waves across the Atlantic, just over a decade after Hertz's assertion that his research was of little practical use. Marconi received the Nobel Prize for his pioneering work on radio transmission in 1909. This is often the way in fundamental physics research; anyone who works at CERN, or NASA, or the European Space Agency, or the European Southern Observatory, or in any field that doesn't produce clearly identifiable widgets, will have been asked to justify the expenditure on curiosity-driven acquisition of knowledge at some stage in their careers. Pointing to the fact that the questioner would probably be dead if a Scottish biologist named Alexander Fleming hadn't isolated penicillin in 1928 because he was curious, rarely does the trick. As Fleming later said, 'When I woke up just after dawn on 28 September 1928, I certainly didn't plan to revolutionise all medicine by discovering the world's first antibiotic, or bacteria killer, but I suppose that was exactly what I did.' How anyone can fail to recognise that understanding the natural world, in which we live and of which we are a part, is unlikely to be useless. Perhaps Fleming could have specified in his will that those who cannot grasp this should be denied the use of his serendipitous discovery? A Darwinian solution to stupidity, admittedly, but evolution by natural selection is also a fact of life. Reason red in tooth and claw.

Einstein also points the way towards the rich insights yet to come as a result of Maxwell's discovery; 'he surely never guessed that the riddling nature of light, apparently so completely solved, would continue to baffle succeeding generations.' As we have already seen in Chapter Two, Einstein felt so strongly about the value of Maxwell's discovery because the universal speed of light was the clue that led him to replace Newton's laws of motion with the Theory of Special Relativity. On its own, this is a most beautiful demonstration of the interconnected character of fundamental physics. Studying electrical currents in wires ultimately mandates a reformulation of our understanding of space and time. But there's much more! F. Scott Fitzgerald said that inserting an exclamation mark is like laughing at your own joke, but I will now attempt to justify its use.

Light as an electromagnetic wave

Light is a wave, according to Maxwell, and it therefore has a wavelength. The wavelength is defined as the distance between two wave-crests, see illustration opposite. Visible light waves are a tiny fraction of the electromagnetic waves travelling through the Universe. They span wavelengths from around 400 nanometres (400 thousand millionths of a metre) in the blue to 700 nanometres in the red. Beyond the red, the electromagnetic spectrum extends to wavelengths too long for our eyes to detect. They are still light – still the sloshing back and forth of electric and magnetic fields driving through the void – it's just that our eyes didn't evolve to see them. Instead we feel them in the residual heat of a fire or the ground at the end of a hot summer's day. Beyond the infrared, we arrive at microwaves, with wavelengths unsurprisingly about the size of a microwave oven. The spectrum then seamlessly slides into the radio region, with wavelengths the size of mountains. For most of our history we have been blind to these more unfamiliar forms of light,

Below: The electromagnetic spectrum.

Bottom: Light as an electromagnetic wave. The wavelength is the distance between two crests.

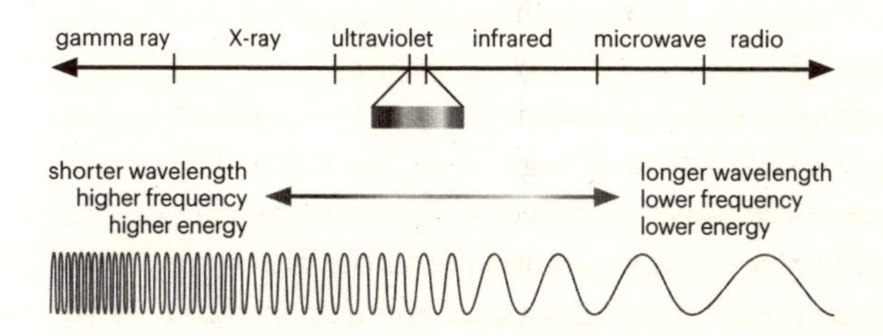

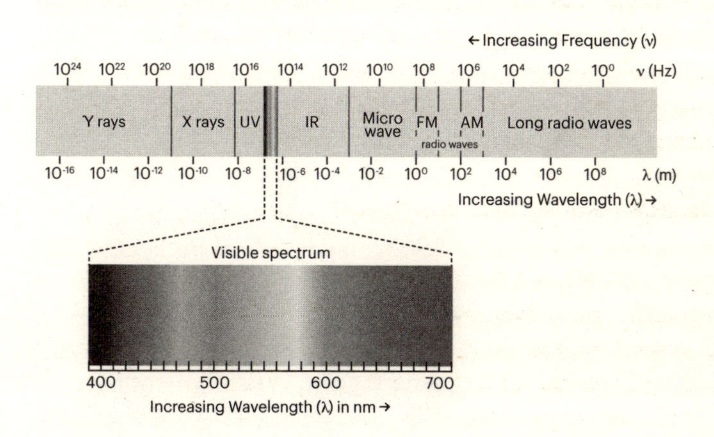

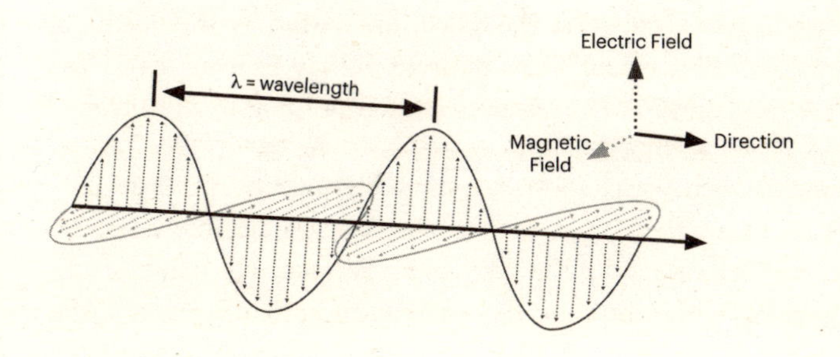

but until recently everyone had a detector capable of intercepting them and turning them into sound. When tuning an old-fashioned radio, you're simply tuning an electronic circuit so that it is sensitive to a particular wavelength of light, broadcast from a transmitter. Music can be encoded in the wave by varying the amplitude of the waves (am radio, standing for amplitude modulation) or the wavelength itself (fm radio, standing for frequency modulation). Today you may be more likely to get your music over the internet, but if you're using wifi, electromagnetic waves are delivering the data, with wavelengths of the order of 10 centimetres.

Just as there is plenty of visible light in the Universe that isn't manmade, so there are also naturally occurring microwaves and radio waves. And, just as for visible light from the most distant galaxies, the microwave and radio light carries information about these distant places across the Universe and into artificial eyes. The sky is ablaze at a wavelength of 21 centimetres, which is the wavelength of light emitted by hydrogen atoms when their solitary electron flips its spin from parallel with the proton to anti-parallel. Telescopes such as the 76-metre Lovell at the University of Manchester's Jodrell Bank Observatory scan the skies at or around these wavelengths.

At shorter wavelengths, beyond the visible, there is ultraviolet light. The Sun glows brightly in the UV, which we cannot see but we feel its effect on our skin as sunburn. At shorter wavelengths there are X-rays, which can penetrate skin just as visible light penetrates glass, but are absorbed by bone, making them useful for medical imaging. Finally, at ultra-short wavelengths, are gamma rays, produced by high-energy astrophysical events such as supernova explosions and in nuclear radioactive decay processes. Gamma ray bursts are some of the highest-energy phenomena in the known Universe; bright flashes of electromagnetic radiation thought to be caused by the deaths of super-massive stars or collisions between binary neutron stars. The brightest gamma ray bursts release energy equivalent to converting a hundred planet Earths into pure radiation.

Why do hot things shine?

Part 2: Max Planck and the Quantum Revolution

We now understand in broad outline that matter emits light because it is made up of moving electrically charged particles. In the language of fields, when electrical charges jiggle they create a changing magnetic field, which creates a changing electrical field, which creates a changing magnetic field, and so on, and the resulting moving disturbance *is* light. Maxwell's equations describe this process mathematically.

This should immediately suggest a link between the temperature of something and the light it emits. The temperature of something is a measure of how fast its constituents are 'jiggling around'; the higher the temperature, the more jiggling, and therefore the 'more light'. We've been deliberately vague here, but the details matter. The correct answer, discovered by the German physicist Max Planck in 1900, saw the introduction of the fundamental physical constant that lies at the heart of quantum theory – Planck's Constant.

Here's why we are allowed an exclamation mark, Fitzgerald be damned. In order to answer the question of how hot things emit light, we've already been led to the door of Einstein's Theory of Special Relativity via Maxwell's equations. We now find that we stand at another door and the other great pillar of twenty-first-century physics, quantum theory, lies beyond. Yet again, we face the interconnectedness of physics. Without an understanding of quantum theory, we wouldn't understand the structure of atoms, possess accurate theories describing the action of three of the four fundamental forces of Nature, or be able to read the stories of distant planets from their reflected light alone. At a more prosaic level, there would be no transistors, and therefore no electronics, and the modern world would be a very different place. Imagine a valve-powered iPhone; it would have a shit battery life.

Planck's foundational insight came to him on the evening of 7 October 1900. We know this because he spent the afternoon at his house in Berlin with a colleague, Heinrich Rubens, discussing theoretical models for the emission of light from hot objects. The experimental results, which were well known and of high precision, are shown schematically in the illustration, below.

The problem with the theoretical models of the day was that they all overestimated the amount of short-wavelength light emitted at a given temperature. Use of the term 'overestimated' might be to understate the problem; the preferred pre-Planckian model, known as the Rayleigh–Jeans law, predicted that an infinite amount of energy should be radiated away at shorter wavelengths by a hot

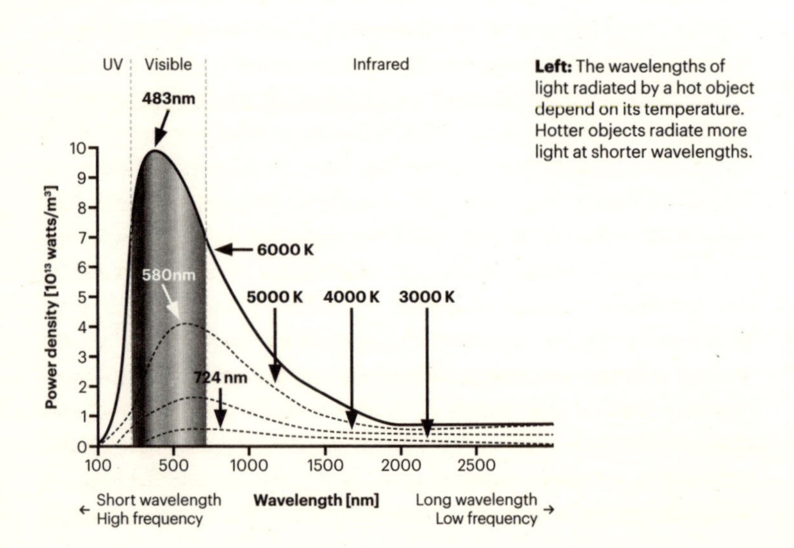

Left: The wavelengths of light radiated by a hot object depend on its temperature. Hotter objects radiate more light at shorter wavelengths.

object. This is obviously not right. The problem lay with the use of one of the foundational theorems of classical physics known as the equipartition theorem. If a lump of matter is considered as a series of little oscillating electric charges that radiate light, in accord with Maxwell's equations, then the equipartition theorem states that all oscillations available to the electrical charges will happen, and they will all share the available energy equally. Faster vibrations correspond to shorter wavelengths of light, and according to classical theory there are more fast vibrations available to the charged particles than slow vibrations. If there is no reason why faster oscillations can't happen, they should dominate and more light should be radiated away at the short-wavelength ultraviolet end of the spectrum, simply because there are more vibrations available. This was known as the Ultraviolet catastrophe, because it is not how hot things behave. Indeed, as can be seen in the illustration opposite, cooler objects don't emit much UV light at all.

When Rubens left the house after a long lunch, Planck was no nearer to a solution, but by the evening he'd sent his friend a formula, scribbled on the back of a postcard. Planck described it as an act of desperation, having tried everything else he could think of. In his scientific biography of Albert Einstein, Abraham Pais writes that Planck's reasoning was 'mad, but his madness has that divine quality that only the greatest transitional figures can bring to science'.

For reasons unknown, not even fully to himself, Planck decided that light can only be emitted in packets, or quanta, whose energy is related to their wavelength through the formula $E = h\,c/\lambda$, where c is the speed of light, λ is the wavelength of the light and h is a completely new constant of Nature which is now known as Planck's Constant. With that assumption he was able to derive the correct description for the spectrum of light emitted by an object of a given temperature. To see how this works, notice that Planck's formula says that shorter wavelengths of light carry more energy, and since there is only a limited amount of energy available, shorter wavelengths will be harder and harder to radiate. The extreme scenario would be for a wavelength that would require more energy than is present in the object: Planck's assumption provides a natural cut-off at the short-wavelength end of the spectrum, and solves the Ultraviolet catastrophe.

Planck thought this was a neat mathematical trick, and didn't appreciate its fundamental physical significance for many years. The reason we quote from a biography of Einstein is that it fell to Einstein, yet again, to take Planck's prediction seriously as a fundamental discovery about Nature. In 1905 he proposed that light is not only emitted and absorbed in little packets, but is actually composed of little packets, called photons. This is not a trivial distinction. Until Einstein, everyone assumed that Planck's insight related to the structure of matter itself, and not to Maxwell's electromagnetic field, which must surely be able to oscillate freely in accord with his beautiful equations. Einstein suggested something much more radical – that the electromagnetic field itself is made up of little particles of light. Just as he replaced Newton's laws with Special Relativity, Einstein proposed that Maxwell's equations are an approximation to something deeper. As late as 1913, Planck was having none of it. In a proposal written in support of Einstein's admission to the Prussian Academy in that year, Planck wrote: 'In sum, one can say that there is hardly one among the great problems in which modern physics is so rich to which Einstein has not made a remarkable contribution. That he may sometimes have missed the target in his speculations, as, for example, in his hypothesis of light quanta, cannot be held too much against him, for it is not possible to introduce really new ideas even in the most exact sciences without sometimes taking a risk.'

Below: A Feynman diagram of an electron emitting a photon, which is absorbed by another electron.

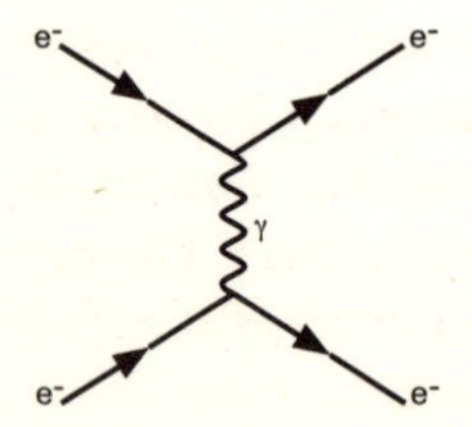

Einstein's instincts, as usual, turned out to be correct. There is a deeper theory than Maxwell's called quantum electrodynamics, which was formulated by Richard Feynman and others during the 1940s and 50s. It was for this theory that Feynman, Julian Schwinger and Sin-Itiro Tomonaga shared the 1965 Nobel Prize in Physics. Einstein himself received the 1921 Nobel Prize for his explanation of something called the photoelectric effect, which was motivated by Planck's insight. Light shining on a metallic surface causes electrons to be released from that surface, but if the light is all above a certain wavelength, no electrons will be released no matter how bright the light. The explanation is that photons of light of too long a wavelength have too little energy to release the electrons, and it doesn't matter whether a million or a billion or a trillion photons hit the metal, no electrons will be emitted because they will never encounter a photon with enough energy to release them. Einstein's explanation is regarded, along with Planck's explanation for the observed spectrum of light emitted by hot objects, as the birth of quantum theory.

We now have everything we need to understand how glowing objects emit light, and why cooler objects emit redder light. Temperature is a measure of how fast things move around, which is a measure of how much energy is available. Electrically charged particles emit light when they are accelerated, in accord with Maxwell's equations. Thinking in this way doesn't explain the colour of the light emitted by hot objects. For that, we need quantum theory. Light can be treated as a stream of particles, whose energy is inversely proportional to the wavelength of the light in accord with Einstein's extension of Planck's hypothesis. Richard Feynman introduced a beautiful way of picturing the process known as a Feynman diagram (see left).

Electrons can emit and absorb photons. The photon will carry away energy and momentum from the electron and deliver it to another one. In this case, we can image one electron being inside a lump of glowing lava. If it's got a lot of energy, it is more likely to emit a photon of high energy, which can be radiated out and absorbed by another electron, which could be inside your retina. This is how you see the world. Since high-energy photons have shorter wavelengths, hotter objects will have a higher probability of emitting short-wavelength photons, simply because the charged particles inside them have more energy on average with which to emit them. Hot things are more likely to emit short-wavelength blue photons, which is why hot things glow blue and cooler things glow red.

We can now round everything off and answer our initial question about why the Sun shines. It shines because its outer layers are jiggling around, heated by the nuclear fusion reactions in its core. The temperature at the surface is approximately 5500 degrees Celsius, and this is a measure of how much energy is available for the charged particles in its surface to emit photons. The solar spectrum is shown in the illustration, below. Because the surface is 5500 degrees Celsius, the peak power is radiated in the visible part of the spectrum. All visible wavelengths are present, which is why the Sun appears 'white hot' in the sky. The surface is hot enough to radiate into the ultraviolet, down to wavelengths of around 250nm, and there is a long tail of emission into the infrared. Planck's theoretical curve, for a perfect emitter (known as a blackbody) of temperature 5500 degrees Celsius, is also shown.

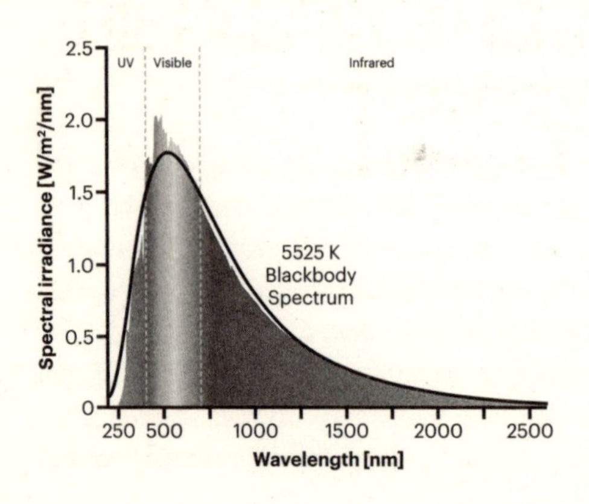

Left: The solar spectrum. Super-imposed is the calculation from Planck's formula showing the spectrum from a 'blackbody' at a temperature of 6000 degrees Celsius.

THE QUESTION OF THE NATURE OF LIGHT AND HOW HOT OBJECTS EMIT IT IS A DEEP ONE. IT REQUIRES A GOOD SELECTION OF THE TOOLS AVAILABLE TO AN EARLY TWENTIETH-CENTURY PHYSICIST.

A serendipitous aside; the solar neutrino problem

The question of the nature of light and how hot objects emit it is a deep one. It requires a good selection of the tools available to an early twentieth-century physicist, in the guise of Maxwell's equations and quantum theory, to answer satisfactorily. If we are also asked to explain how the Sun shines, then we require mid-twentieth-century nuclear physics. The nuclear fusion process we've described was first outlined in detail in a classic theoretical paper, 'Energy Production in Stars', by Hans Bethe in 1939, but experimental confirmation that there are nuclear reactions in the Sun's core came, I find quite astonishingly, in my lifetime.

In 1964 John Bahcall and Raymond Davis Jr proposed an experiment using 100,000 gallons of cleaning fluid to detect the neutrinos produced in the nuclear fusion of hydrogen into helium in the Sun. In an essay written in 2000, Bahcall recalled that the sole motivation of their experiment was to 'see into the interior of a star and thus verify directly the hypothesis of nuclear energy generation in stars'.[1] The first results were published in 1968, and whilst neutrinos were seen, there were fewer than predicted. The number of neutrinos detected was a factor of two or three lower than most refined theoretical models of the Sun suggested. This discrepancy between the observed neutrino flux at the Earth's surface and the predictions from nuclear physics became known as 'the solar neutrino problem'. A series of experiments around the world followed, many throughout my professional career. I remember lecturing an advanced course on neutrino physics in the 1990s in which I presented the solar neutrino problem as one of the unsolved problems in modern physics. Experiments observed neutrinos from the Sun, from cosmic ray collisions high in the Earth's upper atmosphere, and from nuclear reactors. Beams of neutrinos were produced in particle accelerators and angled through the Earth to detectors beneath mountains. The experimental searches were backed up by a great deal of theoretical effort.

The answer to the solar neutrino problem is now known. It came as quite a surprise and has led to one of the most active and exciting areas of research in modern particle physics.

The upshot is that the nuclear physics and the solar models are both correct, but the neutrinos themselves behave in a strange way during their voyage through the Sun and across 93 million miles of space to the Earth. If you look back at the illustration on page 40, you'll see that there are three kinds of neutrino; the electron neutrino, the muon neutrino and the tau neutrino. These are known in the jargon as 'flavours'. Only electron neutrinos are produced in the nuclear reactions in the Sun, and it is the number of electron neutrinos that theoretical physicists calculated and the experimentalists expected to see in their detectors on Earth.

It turns out, however, that Nature is slightly 'misaligned'. Neutrinos don't travel as electron, muon or tau neutrinos, but as a mixture of them. The precise fractions of each that will be detected on Earth depends on the distance they have travelled since they were created, and on what they have travelled through. The early detectors on Earth were only set up to detect electron neutrinos, and they saw fewer than the nuclear physics models predicted – not because there were fewer neutrinos arriving at the Earth from the Sun, but because some of them were arriving as muon or tau neutrinos which escaped detection. This peculiar behaviour is known as neutrino oscillations, and explaining precisely how and why it happens is an unsolved problem. The 2015 Nobel Prize in Physics was awarded to Takaaki Kajita and Arthur B. McDonald for their experimental proof that muon neutrinos created in cosmic ray collisions in the Earth's atmosphere and electron neutrinos created in the Sun's core can transform into the other flavours as they travel from their point of origin to detection.

The interest in the strange behaviour of neutrinos extends way beyond the nuclear physics of the Sun and the behaviour of cosmic rays striking the Earth. In yet another example of the serendipitous twists and turns of science, the discovery of neutrino oscillations has opened up a wonderful can of worms – and cans of worms are a physicist's delight. In order to oscillate in the observed way, at least two of the neutrino types should have very tiny but non-zero masses; around a millionth of the mass of the lightest Standard Model matter particle other than the neutrinos: the electron. We now have good evidence that the Higgs particle is responsible for the masses of the other Standard Model particles, but the enormous difference in mass between the neutrinos and everything else suggests that some other mechanism may be responsible for the neutrino's tiny mass. One such mechanism, known as the see-saw mechanism, requires a new super-heavy neutrino with a mass of the order of 10^{15} GeV; the mass of the proton is approximately 1 GeV. This would be a window into super-high energy physics close to the energies at which it is thought the three non-gravitational forces combine, known as the GUT or Grand Unification scale. Apologies for the units of mass, pronounced G E V or 'Giga electron-volts'. They are more sensible for particle physicists to use than grams. The proton's mass is approximately 1.673×10^{-24} grams, which is an unwieldy quantity. If physicists can use numbers close to 1, they are much happier.

The neutrinos may also have been intimately involved in producing the observed discrepancy between matter and anti-matter in the Universe today – another of the great unsolved mysteries in the physics of the early Universe. Without a difference in behaviour between matter and anti-matter, known as CP-violation, we would not exist. To quote the 2015 Nobel Prize committee, 'the discovery of neutrino oscillations has opened a door towards a more comprehensive understanding of the Universe we live in.'

John Bahcall finishes his lovely essay on the mystery of the neutrinos with this magnificent paragraph:

> *'At the beginning of the twenty-first century, we have learned that solar neutrinos tell us not only about the interior of the Sun, but also something about the nature of neutrinos. No one knows what surprises will be revealed by the new solar neutrino experiments that are currently underway or are planned. The richness and the humour with which Nature has written her mystery, in an international language that can be read by curious people of all nations, is beautiful, awesome, and humbling.'*

[1] http://www.nobelprize.org/nobel_prizes/themes/physics/fusion

Pale blue green planet

Part 1: The Oceans

The white light of the Sun shines onto the Earth and is reflected back out into space. Our planet is an infinitely colourful world close up; cities, jungles, grasslands and savannah have been painted by life – there are few monochrome places. From high altitude, a simpler picture presents itself. The White Marble image (see page 16, plate section) shows an unusual polar view of Earth, centred on Eastern Europe and Russia and extending from the North Pole to the Persian Gulf and India. Four colours dominate in this photo: the blue of the oceans, the green of the temperate northern landmass, the ochre of the deep continental deserts and the white of the clouds and polar snows. What is the origin of these colours, and what can they tell us about the physical and biological processes taking place on the Earth's surface?

In the southwest of Iceland there is a valley called Thingvellir. The first Viking settlers located their parliament in this valley over a thousand years ago, and although parliament was moved in 1798, the site still plays an important role in Icelandic culture. It is a place that symbolises the coming together of a people; a place where information can be exchanged and disputes settled; a place for trade and meeting old friends; a place to work together – a necessary, if not sufficient, requirement for survival on an isolated rock. The valley is narrow and steep in places, which gives the visitor the opportunity to spread out their arms and, almost, touch two of Earth's great continental landmasses. To the east lies North America; to the west, Eurasia. Thingvellir is the only place on land where the mid-Atlantic ridge is visible – a fleeting glimpse of the geological seam that runs the length of the Atlantic Ocean. The ridge is currently spreading at a rate of 2 centimetres per year in the North Atlantic, driving the continents apart. If you've ever wondered why South America looks like it would fit perfectly with Africa, this is the reason. The two continents were one, 130,000 years ago, and the volcanic activity along the mid-Atlantic ridge has carried them quickly apart, the new land of the ocean floor created along a fault line that passes straight through Thingvellir. The Icelandic parliament used to sit at the *Logberg*, or Law Rock, a rocky outcrop that vanished long ago as the landscape shifted. In this part of the world, geology outpaces politics.

There is a place in the great valley that is flooded by crystal glacial meltwater from the central Langiokull glacier, filtered on its journey coastward from the interior through hundreds of kilometres of volcanic rock. The water, transparent and cold, creates one of the world's most famous dive sites.

This is a book inspired by a television programme. Television is a visual medium, and very often the demonstration of some physical principle or other is good for the screen but not for print. However, in our film about the colours of the world, the production team dreamt up a magnificent way of demonstrating why the oceans are blue which works for both media. The sequence involved me diving into the fissure at Silfra (as the site is known) wearing a red dry suit. I imagine the experience is as close as you can get to a spacewalk without actually visiting the International Space Station, because you cannot see the substance of the water; when the sediments settle, it is as if you are unsupported in a silent walled rift. The view from inside the mask was of a blue, enclosed world, but the clue to the origin of the blue light was the red of the suit. On the surface, it was very red. As we descended through 15 metres into the rift, the illumination through the clear waters was still bright, but the suit became black.

The colour of an object is determined by the way light interacts with it. A carrot is orange, for example, because *b-carotene* molecules selectively absorb blue photons. Orange is what's left of the visible spectrum when blue light is removed and, since we see a carrot by its reflected light, it appears orange. Similarly, the dyes in my dry suit absorbed all the colours of the spectrum other than red. The colour from the suit gradually bled away as I descended deeper into the fissure because water molecules absorb red light very strongly. By the time I reached a depth of around 15 metres, there were very

FOUR COLOURS DOMINATE: THE BLUE OF THE OCEANS, THE GREEN OF THE TEMPERATE NORTHERN LANDMASS, THE OCHRE OF THE DEEP CONTINENTAL DESERTS AND WHITE OF THE CLOUDS AND POLAR SNOWS.

few red photons left from the sunlight that entered the water at the surface to reflect off my suit and into the camera lens. The suit continued to absorb all the other colours, which pass through the water relatively unimpeded, which is why the suit turned black, even though illumination levels were still high.

The way water absorbs visible light is quite unique. We saw in Chapter One that water molecules are made up of two hydrogen atoms, bonded to a single oxygen atom. The structure is maintained by the distribution of electrons around and between the atomic nuclei. Electrons can only arrange themselves in very specific ways inside molecules, determined by the laws of quantum theory. Rearrangements can happen without breaking up the molecule, but each different arrangement will, in general, have a different energy. If the arrangement of electrons inside a molecule is to be changed, a photon with just the right energy to make the change must be absorbed. Since the energy of a photon is directly related to its colour, a particular molecule will only absorb certain colours of light, determined by the different possible arrangements of electrons inside it.

Below: The three basic modes of vibration of a water molecule.

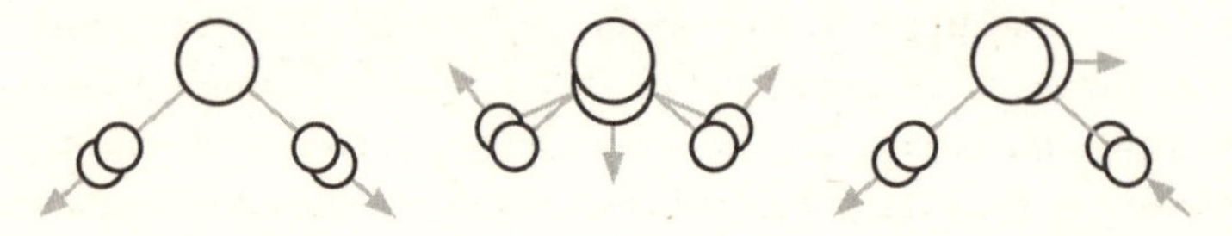

This is the process by which virtually everything we see acquires its visible colour; but water is different. The arrangement of the electrons inside water molecules does change as a result of the absorption of electromagnetic radiation, but the energies required are too high for photons in the visible part of the spectrum to be involved. Instead, it is vibrations between the hydrogen and oxygen nuclei inside the water molecules themselves that are driven by the absorption of lower-energy infrared and visible (red) photons.

There are three basic modes of vibration of a water molecule, shown in the opposite illustration, but a tremendous array of combinations is possible, leading to water's extremely complex absorption spectrum – which is shown in the graph on page 232. Many of these vibrations are excited by long-wavelength infrared photons, and this is the mechanism exploited in a microwave oven. There are also vibrations that can be excited by visible red light, removing it from the spectrum. Water is virtually opaque to ultraviolet light, and to infrared light, which is where the intra-nuclear vibrations kick in. But there is a valley, mainly in the blue and green, where water does not absorb light strongly. This is why water looks 'almost' transparent. The steep rise in absorption towards the red part of the visible spectrum is the reason why my red dry suit lost its colour. At a depth of 15 metres, the intra-nuclear vibrations of the water molecules have absorbed most of the red photons from the Sun that entered the surface, and there are few left to be reflected by the dry suit. In deeper water, all that is left is blue light, which is scattered around rather than absorbed.

This is what gives large bodies of liquid water their planet-defining blue hue. We are a 'Pale Blue Dot' because of the delicate interaction between electromagnetic radiation and the rotating, wobbling, vibrating molecules formed by the first and third most common elements in the Universe: hydrogen and oxygen.

As an aside, it's interesting to note the sensitivity of the absorption spectra of molecules to slight changes in their constituents. Heavy water is chemically identical to H_2O, but it contains deuterium rather than hydrogen. It has the chemical formula D_2O. Deuterium is an isotope of hydrogen, and its nucleus contains a single neutron alongside the proton. This has no effect on the chemistry, which is driven purely by the number of electrons that surround the nucleus and therefore the number of protons inside it. The physical presence of the neutron does have a very noticeable effect on the absorption spectrum, however. Instead of absorbing light in the red part of the visible spectrum, the vibrational modes are shifted to higher energies, and therefore excited by shorter-wavelength photons beyond the visible. This is in accord with intuition; it takes more energy to make a more massive nucleus vibrate back and forth. As a result, since virtually none of the visible spectrum is removed, heavy water is colourless even in large quantities. If the Earth were covered in oceans of D_2O, it would not be a blue planet.

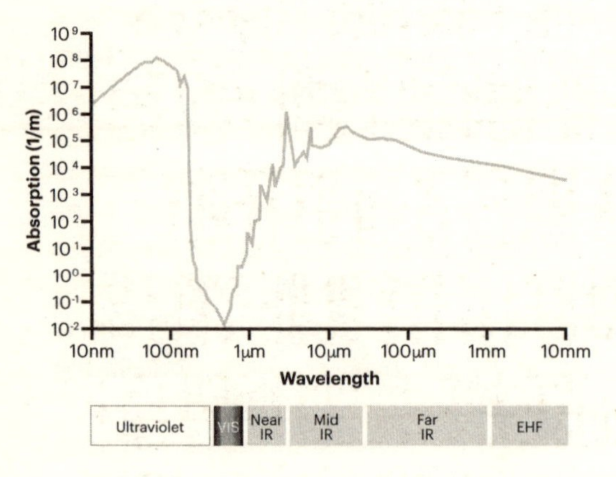

Left: The absorption spectrum of liquid water.

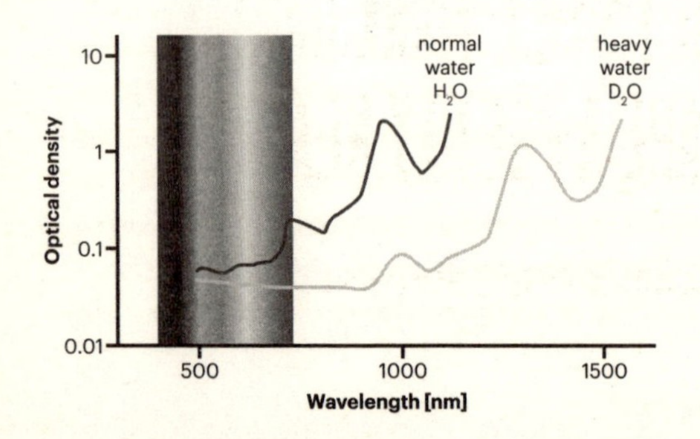

Left: Comparison of absorption of H_2O and D_2O.

‘FAR OUT IN THE UNCHARTED BACKWATERS OF THE UNFASHIONABLE END OF THE WESTERN SPIRAL ARM OF THE GALAXY LIES A SMALL UNREGARDED YELLOW SUN. ORBITING THIS AT A DISTANCE OF ROUGHLY NINETY-TWO MILLION MILES IS AN UTTERLY INSIGNIFICANT LITTLE BLUE-GREEN PLANET WHOSE APE-DESCENDED LIFE FORMS ARE SO AMAZINGLY PRIMITIVE THAT THEY STILL THINK DIGITAL WATCHES ARE A PRETTY NEAT IDEA.’

– DOUGLAS ADAMS, THE HITCHHIKER’S GUIDE TO THE GALAXY

Pale blue green planet

Part 2: The Sky

The Earth's blue skies are not the result of the selective absorption of sunlight, but of selective scattering. Here again, the demonstration cooked up by the television production team is rather instructive. On the crisp Monday morning of 28 September 2015, a total lunar eclipse was visible from the UK. Every two and a half years or so, the Sun, Moon and Earth align such that the Earth, positioned briefly between our star and satellite, casts a shadow across the face of the Moon. It is a beautiful and relatively common sight, and one that delivers a powerful component of the feeling I experienced when I watched a total solar eclipse from Varanasi, in India, in 2009. Both are a display of moving shadows, cast across the Solar System by orbiting balls of rock, and once you have that in your mind, the effect of an eclipse is all the more powerful. During a lunar eclipse, the shadow of our world is visible, and it is lonely and dark against the Moon. In Varanasi, by the banks of the Ganges on a dripping tropical July morning, heavy with sweet incense and sweat, a million voices fell silent as the shadow of the Moon darkened the magical old Ghats. In England, thousands of miles, seven long years and a great spiritual rift away, as I prepared to recount my feelings on a quiet English moor, two 'witches' decided to mark the occasion by singing the theme from Walt Disney's *Frozen*.

The Earth's shadow completely covers the Moon during a total lunar eclipse, but the Moon doesn't fall into absolute darkness. Instead, it glows a dim, deep red. The red illumination of the lunar surface is the result of sunlight being deflected onto the Moon by Earth's atmosphere. The Moon is normally viewed in direct sunlight; it reflects 12 per cent of the visible spectrum – a little less from the dark basalt seas laid down by ancient volcanic eruptions, and a little more from the brighter anorthosite highlands. Asked to describe moonlight in a single word, you'd probably say white; not too different from sunlight. This is because Moon rocks reflect light reasonably democratically at all wavelengths; bright rainbow in – dimmer rainbow out. There are certainly no reds, greens and blues visible to the naked eye. During an eclipse, the illumination is very different. The Earth's atmosphere acts as a filter, removing most of the solar spectrum other than the red light, which remains to illuminate the maria and highlands. This is why the Moon turns red during a lunar eclipse.

The same physical process turns the sky red at sunset. As the Sun falls, or should we say as the Earth rotates beneath the Sun, the sunlight has to travel through an increasing amount of atmosphere on the way to our eyes. The image of the Sun reddens, and as the Sun approaches the horizon, the sky itself turns from blue to red. To understand what is happening, we need to know how photons of different wavelengths, and therefore energies, interact with the molecules, dust and water vapour in the Earth's atmosphere.

Below: The red illumination of the lunar surface occurs when sunlight is deflected onto the Moon by Earth's atmosphere.

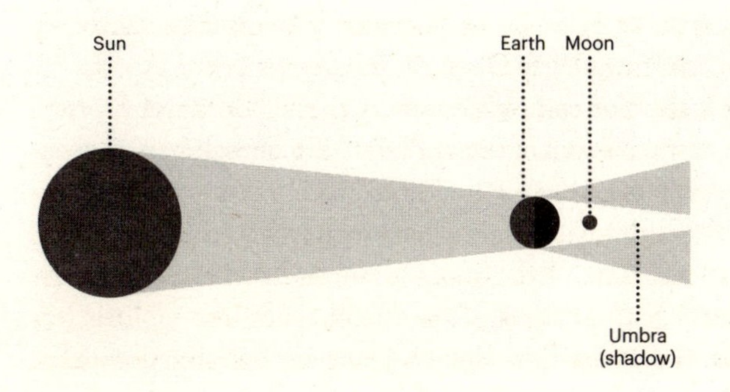

The changing colours of Earth's skies, and the deep red of the lunar surface during an eclipse, are caused by a process known as Rayleigh scattering, named after the British physicist Lord Rayleigh (John William Strutt). The process can be described as the elastic scattering of photons off the oxygen and nitrogen molecules that make up our atmosphere. Picture billiard balls bouncing off each other; this is a good image if the wavelength of the incoming light is significantly larger than the size of the molecules, which is the case for visible light making its way through the air. The wavelengths of visible photons are between 400 and 650nm, and oxygen and nitrogen molecules are over a thousand times smaller.

In modern language, Rayleigh's formula shows that the probability for a photon to scatter is inversely proportional to the fourth power of its wavelength. This means that blue photons (450nm) are over three times more likely to scatter off gas molecules on their way through the atmosphere than longer-wavelength red photons (650nm). The illustration opposite shows the percentage of sunlight that is scattered on its way through the atmosphere when the Sun is directly overhead. Around one in five blue photons scatter, whereas only one in twenty red photons will be deviated from a straight-line path from the Sun into your eye. This is why the sky appears blue and the Sun takes on a yellow tinge. As the Sun drops towards the horizon and the photons have to journey through more air, the chance of any photon scattering will increase, and in particular more of the blue light is scattered away. This is why the skies become increasingly orange and even red in the evening, leaving a fading, deepening disc of red as the Sun falls below the horizon.

Thanks to the Apollo astronauts, we can see what the Sun looks like in a sky with little or no atmosphere. Photographs taken on 19 November 1969 by the team of astronauts on board Apollo 12 showed us that the Sun is bright white over the 'Ocean of Storms' because none of the colours of the rainbow have been scattered away and the sky is deep black.

From Earth's orbit our atmosphere is rarely visible, although photographs of the limb of the Earth from the International Space Station provide a dramatic view of the thin blue line that separates us from the vacuum of space (see page 16, plate section). The dominant atmospheric features that are visible from space are the bright white clouds. Clouds are white because they are composed of water droplets, which are typically of comparable size to the wavelength of visible light. Rayleigh's calculation does not apply here, and the dominant scattering process is known as Mie scattering, after the German physicist Gustav Mie. Larger particles, such as water droplets, scatter light with a probability that is almost independent of wavelength, and this democratic deflection is the reason why clouds on Earth are bright white.

Below: The percentage of sunlight scattered by the Earth's atmosphere when the Sun is directly overhead, as a function of wavelength.

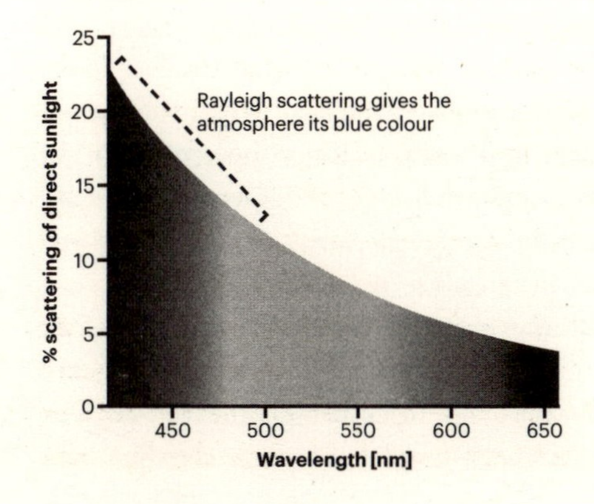

Pale blue green planet

Part 3: The Land

Beneath the white clouds, lined by the blue oceans, is the land. The polar regions are white, the equatorial belts a dusty Mars-red, but the temperate north on the White Marble image is green. As I write, looking at that photograph (see page 16, plate section), I'm taken aback by just how green Europe and northern Asia are. There is no sign of concrete or highways or cities. The surfaces of Britain, France, Germany, the lowlands of Norway and Finland and out across the eastern planes of Russia, halfway around the globe to the North Pacific coast, are uniformly verdant. The ring of green is completed by North America, just visible through the clouds off the upper limb. These are the places where we know there will be abundant food, shelter and rain, because we recognise green as the colour of life. But why are plants green?

As with many of the simply phrased questions we've asked in this book, there are multiple answers to this of increasing depth, and at the end of the thread lies that most wonderful answer for a scientist: we don't quite know yet. With that exciting and tantalising admission of ignorance, a most magnificent and humble thing, let's start with what we do know.

A very simple answer is that green is the colour that life throws away. Just as the oceans are blue because water molecules do not readily absorb blue photons, so plants are green because chlorophyll, the pigment contained within all green plants, absorbs blue and red photons and the green photons are reflected back out into our eyes. The diagram, right, shows the absorption spectra for the two most common forms of chlorophyll, labelled *a* and *b*. They absorb wavelengths at both ends of the visible spectrum, but leave the green centre well alone. To make more progress in understanding why this might be the case, we need to know a little about the complex biological magic of photosynthesis.

The biochemist Albert Szent-Gyorgi once observed, 'Life is nothing but an electron looking for a place to rest.' Photosynthesis is the process by which plants use energy from the Sun to move electrons around, and today it lies at the heart of the entire food chain. You'll probably remember this basic equation from school:

$$6CO_2 + 6H_2O \rightarrow C_6H_{12}O_6 + 6O_2$$

Strictly speaking, we should refer to this as oxygenic photosynthesis, because the source of the electrons in this case is water, which falls apart, releasing oxygen into the atmosphere as a waste product. Ripping electrons off water is extremely difficult to do. There is perhaps another corner of your mind where, amongst the windmills, you've filed away a school science experiment; the electrolysis of water. Water can be split into hydrogen and oxygen by passing electricity through it, but it's not easy because water is a very stable and tightly bound molecule. If there was an energy-efficient way of splitting water using our current technology, the world economy would be based on hydrogen rather than oil.

Below: The absorption spectra of the two closely related photosynthetic pigments, chlorophyll a and b.

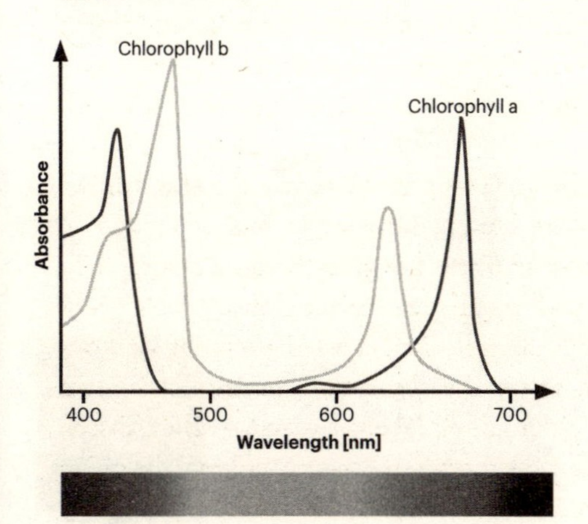

Photosynthesis has been around for a very long time, and probably dates back at least 3.5 billion years to some of the oldest organisms, known as cyanobacteria. These early forms of life would not have possessed the advanced biochemical machinery necessary to split water, and would have grabbed their electrons off less-stable molecules such as hydrogen sulphide, readily available in the oceans of the young Earth. Just as in plants today, they would have forced those electrons onto carbon dioxide to make sugars, the building blocks of living things. They also had the ability to use the electrons liberated by sunlight to manufacture ATP, life's universal battery. At some point earlier than 2.5 billion years ago, an evolutionary innovation known as the oxygen evolving complex allowed organisms to replace hydrogen sulphide with the more readily available water, and the whole lot was linked together to form the Z-scheme, which is present in all green plants today.

The Z-scheme is one of the wonders of evolutionary biology. The sugar-manufacturing piece alone, known as photosystem 1, consists of 46,630 atoms. The ATP piece is known as photosystem 2. The oxygen evolving complex has such an intricate structure that it was not fully understood until 2006.

Below: Z-scheme

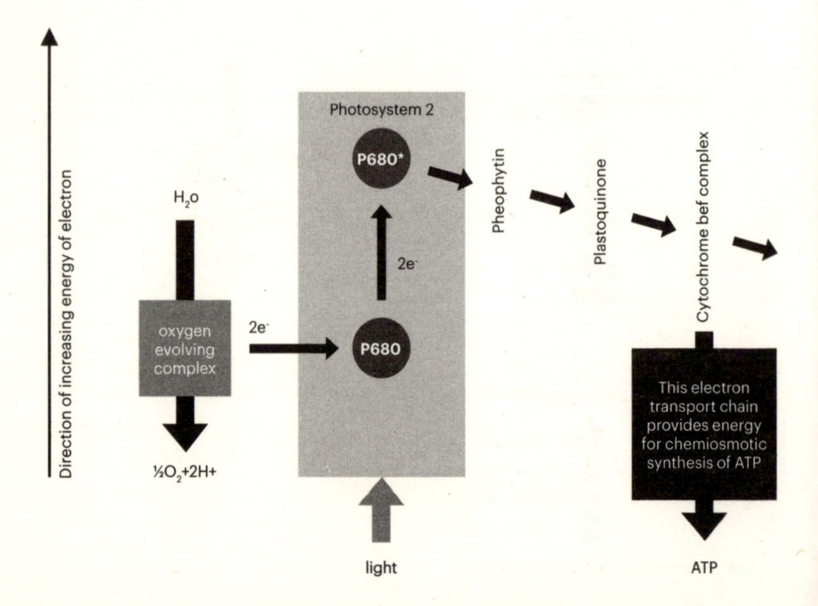

The power source for all this machinery is the plentiful stream of photons from the Sun, and chlorophyll is the primary collector of photons. There are several types of chlorophyll, which perform different functions that depend on their molecular structure and their surrounding proteins. At the reactive heart of photosystem 2, chlorophyll absorbs light most strongly at a wavelength of 680nm, which is in the red part of the spectrum. The energy absorbed reconfigures the distribution of electrons in the molecular structure, resulting in one being made available to the first electron transport chain of the Z-scheme, which whisks it away to manufacture ATP. This leaves the chlorophyll with a voracious appetite to regain its lost electron, which it grabs from water with the help of the oxygen evolving complex. The structure that contains the chlorophyll molecules is known as the P680 reaction centre, and when it has absorbed a photon it is the strongest-known biological oxidising agent. This is why it has the power to split water, delivering the oxygen we breathe into the Earth's atmosphere in the process.

After progressing through photosystem 2, the electron is ready to enter photosystem 1, the business end of which contains another set of chlorophyll molecules inside a different structure called the P700

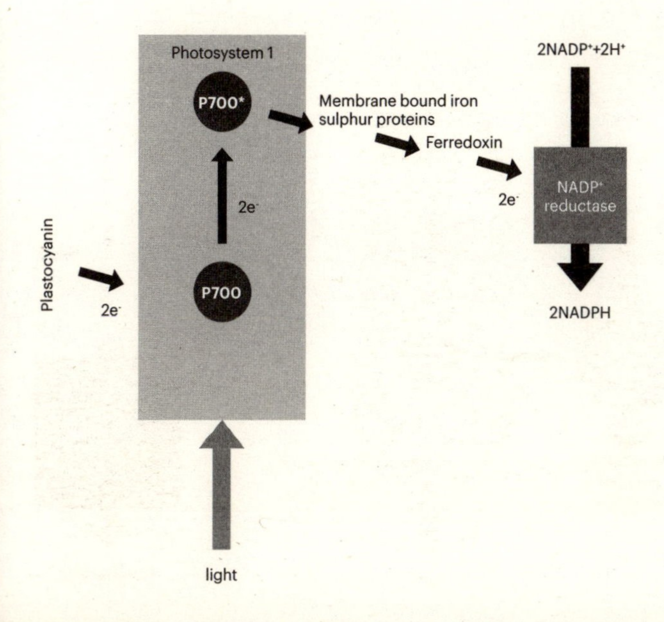

reaction centre. It absorbs light most strongly at the slightly higher wavelength of 700nm, deeper into the red. In this guise, chlorophyll absorbs light just as before, but with a different result. It now becomes the most powerful known biological reducing agent, which means that its appetite is focused on getting rid of its energised electron onto anything it can – in this case, via a few more pieces of molecular machinery, onto carbon dioxide. The result, with the addition of a few protons, is to turn CO_2 into sugars. The missing electron is replaced by the spare one that popped out of photosystem 2.

This may seem unnecessarily complicated, but it probably isn't. If you gave a chemical engineer the job of pulling electrons off water and putting them onto carbon dioxide, she'd probably laugh in your face. Water doesn't want to give up electrons, and carbon dioxide doesn't want to receive them. The job of pulling electrons off a stable thing is very different to the job of putting electrons onto a stable thing, and this is why there are two separate reaction centres allowing the chlorophyll pigments to perform these different tasks.

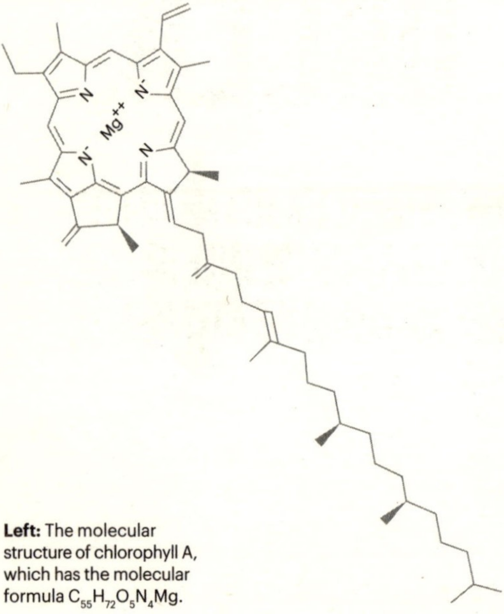

Left: The molecular structure of chlorophyll A, which has the molecular formula $C_{55}H_{72}O_5N_4Mg$.

'A TEARDROP OF GREEN.'

– RON MCNAIR, PHYSICIST AND NASA ASTRONAUT,
ON VIEWING THE EARTH FROM THE SPACE SHUTTLE,
NEWSWEEK MAGAZINE, 10 FEBRUARY 1986

The Z-scheme is an awesome thing, which is probably why every organism on the planet that carries out oxygenic photosynthesis does it in precisely the same way. It almost certainly only evolved once, probably in a cyanobacterium somewhere in a primordial ocean. These clever cyanobacteria somehow found their way into the cells of other organisms and became the chloroplasts – the seat of photosynthesis in all the green plants on the planet today. This may give you pause for thought, because without the Z-scheme there would be very little oxygen in our atmosphere and complex life on Earth wouldn't exist.

If the two reaction centres absorb light most strongly in the red, then why are all plants green? The answer is that the P700 and P680 reaction centres don't absorb sunlight directly. This is done by a complex array of different chlorophyll pigments, and other pigments called accessory pigments, which channel the light into the reaction centres in a cascade that gradually increases the wavelength towards the red part of the spectrum, allowing the chemical business to begin. The accessory pigments are revealed in the autumn, when the chlorophyll decays away, as the reds, oranges and golden yellows of autumn leaves. The two most common chlorophyll pigments outside of the reaction centres absorb light in both the red and blue parts of the spectrum. Together with the accessory pigments, they harvest over 90 per cent of the Sun's light, leaving only a very small band of green to be reflected away.

Photosynthesis is complicated and wonderful. It uses almost all of the sunlight falling on the surface of the Earth to power the plants that lie at the base of our planet's food chain, and oxygenates the atmosphere in the process. Why don't plants use 100 per cent of the visible spectrum and have black leaves, rather than reflecting 10 per cent of the light away? Nobody knows. The answer is probably an important lesson in evolutionary biology. Evolution by natural selection doesn't find optimal engineering solutions to problems. If an engineer designed a plant, it would have black leaves. Rather, organisms are a bit of a bodge job, the result of 4 billion years of mutations, selection pressures and genetic and physical mergers. The greens that dominate the temperate regions of planet Earth could well be a frozen evolutionary accident.

Pale coloured dots

Having taken a wander through the origin of Earth's defining colours, we can now return to the beginning and cast our minds out towards the stars. Is there any way we can use what we know about the reflection and absorption of sunlight on Earth to explore other worlds, and to search for the signatures of life beyond our Solar System? The answer is yes, and astronomers are doing just that.

The first planet to be discovered outside the Solar System is known as PSR B1257+12 B. The discovery was announced in January 1992. PSR stands for pulsar – a rapidly rotating neutron star around 1.5 times the mass of the Sun but with the radius of a city. Pulsars rotate extremely fast – the parent star of the first planet spins around once every 0.006219 seconds. The timing accuracy is important, because it is by measuring wobbles in the spin rate that the existence of planets can be inferred.

There are three known planets in the PSR B1257 system, which have been named Draugr, Poltergeist and Phobetor. Poltergeist was the first to be discovered. I know; I was curious about their names as well. Poltergeist means pounding ghost, the Draugr are the undead in Norse legend who live in their graves, and Phobetor is the personification of nightmares and the son of Nyx, Greek goddess of the night. Astronomers are such Goths. Mind you, the PSR B1257 system wouldn't be a very nice place to live – the planets are bathed in radiation from their violent host. Draugr is the closest in, orbiting once every 25.262 Earth days. It is the lowest-mass planet yet to be discovered, at only twice the mass of our moon.

The Kepler Space Telescope was launched on 7 March 2009 and has revolutionised the search for extra-solar planets. Kepler looks for periodic dips in the light of stars as planets pass across their face as seen from Earth. By studying the details of the light drop, and with additional data from supporting observations by ground-based telescopes, a great deal of information about the planets can be deduced. I write on 11 May 2016, a day after the discovery of 1284 new planets was announced by the Kepler team. In this new sample alone, there are 550 rocky Earth-like planetary candidates, and nine of these orbit in the so-called habitable zone around their parent stars, which allows them to have surface conditions compatible with the existence of lakes and oceans. The 21 rocky planets less than twice the size of Earth discovered by Kepler are shown in the illustration, right.

The Kepler and ground-based data allow for the size, mass and orbital parameters of the planets to be measured, which can be used to estimate their density and temperature and gives a guide to their composition. To go further, starlight that has interacted with the planetary atmosphere itself must be analysed directly, and this can be done.

The first atmospheric analysis of a large rocky planet was reported in February 2016 by a team from University College London, using data from the Hubble Space Telescope.[2] The planet, called 55 Cancri e, is one of five known worlds that orbit around the yellow dwarf star 55 Cancri A, only 40 light years from Earth. The star also has a smaller red dwarf companion, 55 Cancri B. The planet is around 8 times the mass of the Earth, and has an atmosphere of hydrogen and helium. No water vapour was detected, but there were hints of hydrogen cyanide, which astronomers believe indicates a carbon-rich atmosphere. This world is an exotic, violent place, with a year that lasts 18 hours and surface temperatures in excess of 2000 degrees Celsius. It is clearly not a world where we would expect to find life. The significance of the measurement is in the successful retrieval of the vanishingly faint spectrum of a small, rocky planet from the bright, overwhelming light of its parent star.

The direct observation of the light from exoplanets is still in its infancy, but the James Webb Space Telescope, due for launch in October 2018, will allow planetary atmospheres to be probed in unprecedented detail. Kepler's successor, the Transiting Exoplanet Survey Satellite, will be launched in 2017 and will add huge numbers of Earth-like worlds for the JWST to observe, including Earth-sized planets around red dwarf stars. The discovery of water vapour on such a world would be exciting. The discovery of high oxygen levels would be a smoking gun for the presence of photosynthetic organisms. We may be very close indeed to discovering that we are not alone in the Universe.

Would that matter? These planets are beyond physical reach, at least for the foreseeable future, and it is extremely unlikely, in my view, that these planets will be populated by intelligent beings. If life is present, I would guess that it would be microbial. But I could be wrong. In any case, of course it matters. The lights in the night sky are powerful, majestic, but impersonal. The detailed knowledge of a thousand worlds of ice and snow and fire won't, I regret, help us to live better lives – the folly of human conceits is too deeply engrained for that. I believe we will need a collective shock if we are to 'deal more kindly with one another, and to preserve and cherish the pale blue dot'. The shock could be something negative. Perhaps we'll have to come together to fix the climate we're mangling, or deflect a doomsday asteroid. Or, it might be something positive. Astronomy turns data into dreams; if we discover that life is common across the Universe, will it still be possible to glance up at those bright old stars and not feel as one nation beneath them? Why study rainbows? Then we'd know the answer.

Right: Potentially Earth-like planets in the habitable zone around stars discovered by the Kepler Space Telescope (as of May 2016).

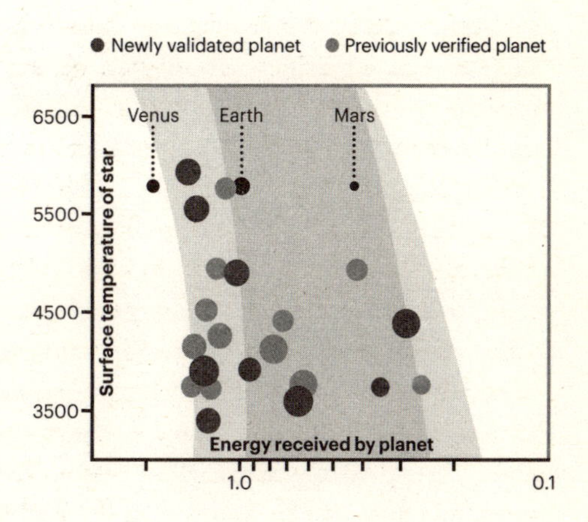

[2] http://arxiv.org/abs/1511.08901

'WHY ARE THERE SO MANY SONGS ABOUT RAINBOWS
AND WHAT'S ON THE OTHER SIDE
RAINBOWS ARE VISIONS
BUT ONLY ILLUSIONS
AND RAINBOWS HAVE NOTHING TO HIDE
SO WE'VE BEEN TOLD
AND SOME CHOOSE TO BELIEVE IT
I KNOW THEY'RE WRONG, WAIT AND SEE
SOME DAY WE'LL FIND IT
THE RAINBOW CONNECTION
THE LOVERS, THE DREAMERS, AND ME.'

– KERMIT THE FROG, THE MUPPET MOVIE

p101 © Print Collector/Getty Images; pp128, 131 © Brian Cox; p150 © Wellcome Library, London. Plate section: pp1, 9 (bottom), 11 © Brian Cox; p2 (top) © Wilson Bentley Digital Archives of the Jericho Historical Society/snowflakebentley.com; p2 (bottom) © NOAA/Science Photo Library; p3 © Brian J. Skerry/Getty Images; pp4–5 © Jose Fuste Raga/Getty Images; p6 © Mondadori Portfolio/Getty Images; p7 (top) © NOAA Archives; p7 (bottom) © NASA/Science Photo Library; p8 © Carrie Vonderhaar/Ocean Futures Society/National Geographic Creative; p9 (top) © Sylvester Chong/Getty Images; p10 (top) © Dirk Wiersxwma/Science Photo Library; p10 (bottom) © BBC; p12 (top) © Arabic Manuscripts Collection/New York Public Library/Science Photo Library; p12 (bottom) © NASA/JPL/Space Science Institute/Science Photo Library; p13 (top) © David Parker/Science Photo Library; p13 (bottom) © NASA/SDO/Getty Images; p14 © The Asahi Shimbun/Getty Images; p15 © Science & Society Picture Library/Getty Images; p16 (both images) © NASA.

Acknowledgements

Brian Cox and Andrew Cohen
May 2016

Forces of Nature has been a long time in the making. Since January 2014 a world-class team led by Danielle Peck and Giselle Corbett have shown huge intellectual and creative ambition and a large dose of stoicism to deliver the television series that accompanies this book. We'd like to thank everyone involved in the production for the endless commitment they have given to the project.

We'd particularly like to thank Matthew Dyas and Stephen Cooter for sticking with it and producing such beautifully crafted and thoughtful films. They were supported by a hugely talented team who have grappled week after week with the demands of production on this scale. So a very big thank you to Alex Ranken, Alice Jones, Suzy Boyles, Duncan Singh, Helene Ganichaud, Mags Lightbody, Francesca Bassett, Emma Chapman, Louisa Reid, Robert Hanger, Wendy Clarke, Laura Stevens, Rebecca Hickie, Nik Sopwith, Simon de Glanville, Julius Brighton, Tim Cragg, Paul O'Callaghan, Graeme Dawson, Adam Finch, Lee Sutton, Damien Sung, Andy Paddon, Paul Thompson, Benji Merrison, Vicky Edgar and Marie O'Donnell.

We'd also like to add a particular thanks to Darren Jonusus who has once again set new standards of craft and creativity in the edit.

Also a huge thank you to Laura Davey the Production Executive of BBC Science who has seen this production through many difficult days in her usual calm and supportive manner.

We'd also like to thank Professor Jeff Forshaw and Professor Matt Cobb for the generous time and thought they have given to the project.

The team at HarperCollins have once again shown their patience and brilliance in equal measure. Holding their nerve in the face of daunting deadlines and a book without words, they have delivered these beautiful pages seemingly overnight. We'd like to thank Zoë Bather, Julia Koppitz, Helena Caldon, Madeleine Penny and of course the very brilliant Myles Archibald (who must spend less time in the dentist's waiting room).

Andrew would like to thank Anna…again…for her love, support and patience through the many late nights.

I would like to thank The University of Manchester and The Royal Society for allowing me the time to make *Forces of Nature*. I would also like to thank Sue Rider, my endlessly wise and supportive agent and friend.

Page references in *italics* indicate diagrams and illustrations.

A

B

C

D

N

Q

R

T

U

V

W

X

Y

Z